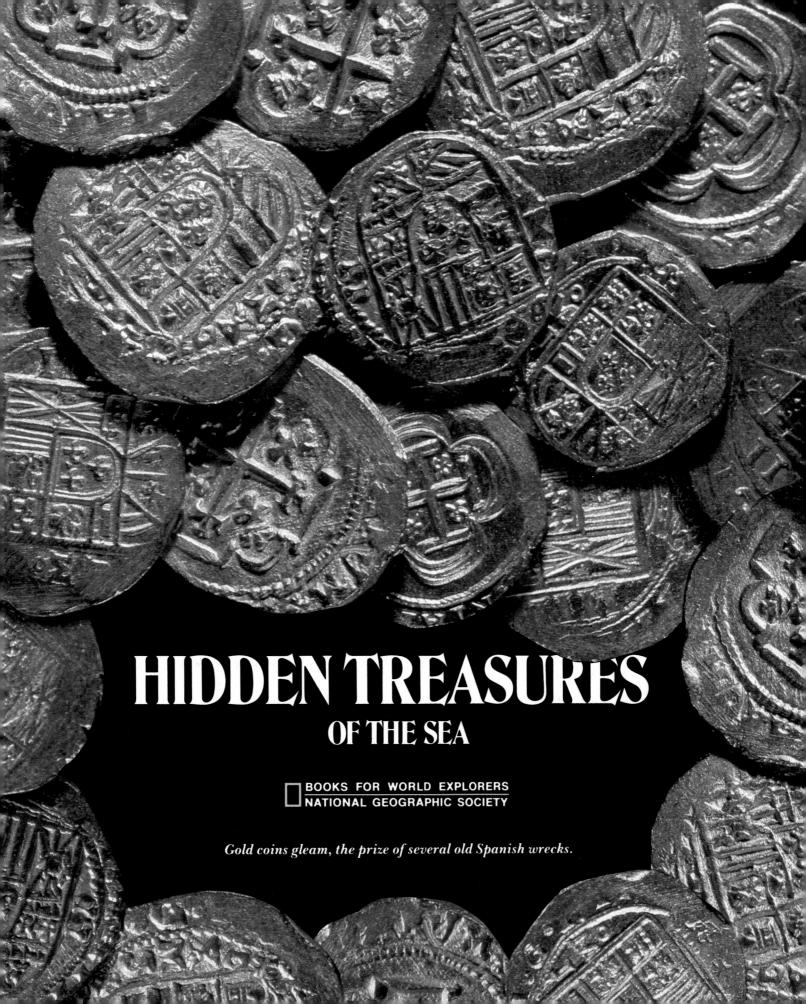

HIDDEN TREASURES
OF THE SEA

☐ BOOKS FOR WORLD EXPLORERS
☐ NATIONAL GEOGRAPHIC SOCIETY

Gold coins gleam, the prize of several old Spanish wrecks.

Contents

1

Voyagers of the Ancient World

GOLD
PENDANT
FROM ANCIENT
WRECK

8

2

Dragonships and Sea Traders

INCENSE
BURNER,
FOUND OFF
SOUTH KOREA

25

3

Whalers and Men-of-War

PART OF
A LINSTOCK,
USED TO
FIRE CANNON

39

BARBER'S
BOWL FROM
SPANISH
WRECK

PART OF A
RIVERBOAT'S
CARGO

Copyright © 1988 National Geographic Society
Library of Congress CIP data: page 103

LANTERN
FROM SUNKEN
JAPANESE SHIP

Diving Into The Past

For thousands of years, ships large and small have traveled the earth's waterways. They have carried wine jugs and warriors, gold and slaves, automobiles and tourists. But not all ships have made it to their destinations. Some have gone down, often taking with them all that they carried.

Sunken ships can hold treasure of various kinds. Treasure may be rubies and emeralds and gold coins. It may be clues to the way people lived in the past. Often it is both. Only in fairly recent times have we come to realize the value of shipwrecks for peering into the past. The key to learning from shipwrecks lies in the careful study of the sites and their contents. That work falls to experts—and in particular to a kind of scientist called a nautical archaeologist.

You can consider an archaeological site a time machine that takes you to a certain period in history. Man-made objects found at a site are called artifacts—and they provide fascinating glimpses into the past. On the following pages, you'll find such artifacts as jewels and gold chains as well as children's building blocks, jars of candied peaches, and a bathtub that looks like a birdbath.

Special equipment is often needed in exploring a wreck. By today's standards, some early diving devices seem crude (left). Still, they enabled divers to go deeper and stay longer than a single gulp of air would allow. Modern-day exploration took a great leap forward in 1943 with the invention of the Aqua-Lung, now commonly called scuba gear. Such gear (pages 6–7) enables divers to work beneath the sea, unconnected to the surface. Divers also use underwater vehicles such as submersibles (right) and remote-control craft. In 1986, scientists used both types of craft (pages 90–91) to explore one of the most remarkable finds ever: the remains of the sunken luxury liner *Titanic*.

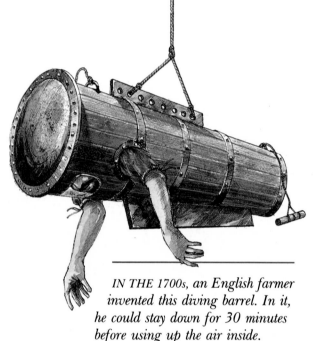

IN THE 1700s, an English farmer invented this diving barrel. In it, he could stay down for 30 minutes before using up the air inside. Today, robots (below) do many diving chores. They need no air at all.

SWIMMING EYEBALL. Kyra Kristof, of Washington, D. C., hoses off a robot named SeaROVER. The remote-control craft takes underwater pictures. Kyra, 8, is helping her father, Emory, on an assignment. He's a photographer with the National Geographic Society.

SEA HELICOPTER. The Johnson-Sea-Link carries as many as four people 3,000 feet down. Scientists have used this submersible to explore the wreck of the Civil War ship U.S.S. Monitor *off the North Carolina coast. Like a helicopter, it can go straight up and down and can hover in place.*

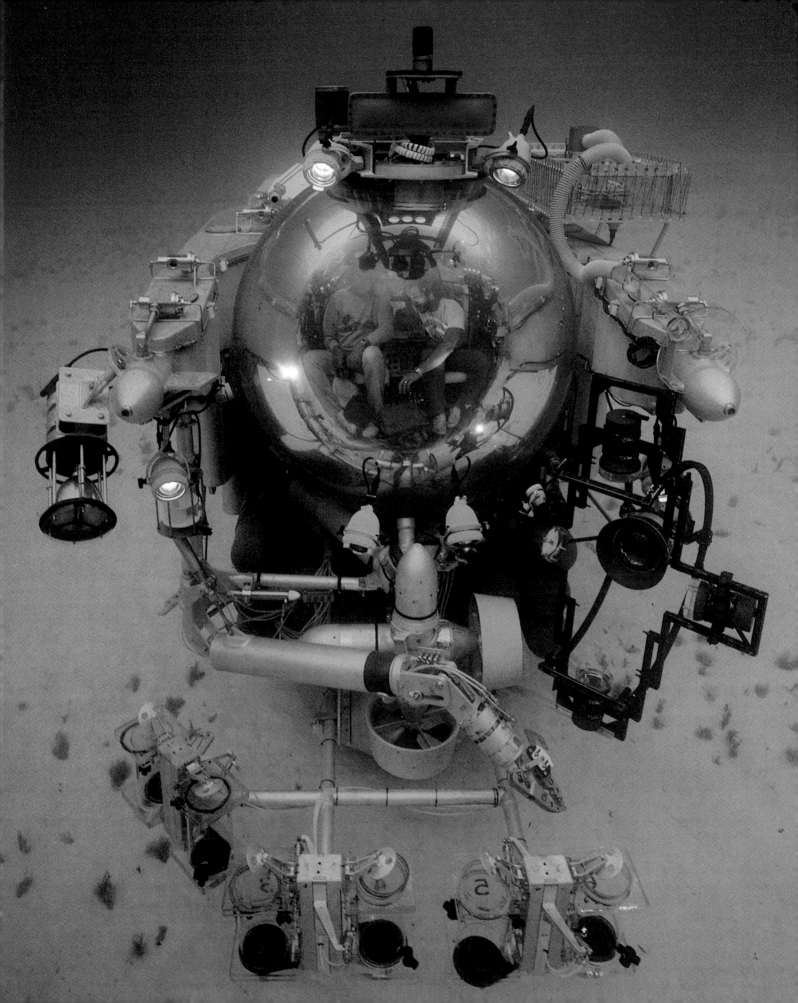

SLOW, CAREFUL WORK helps nautical archaeologists learn as much as they can about a wreck. In this painting, scientists a hundred feet down uncover an ancient ship in the Mediterranean Sea.

Breathing with scuba gear, archaeologists plot a find. A metal grid laid over the site helps them record the location of ship's timbers and other artifacts— objects made by people.

Using a special camera, a scientist takes three-dimensional photographs that will help the team measure the grid area.

Archaeologists make complete records of the location of all artifacts. The information will help them later as they piece together the ship's history.

After details have been fully recorded, delicate artifacts are lifted to the surface by means of a balloon called a "consta lift." Air bubbling from the balloon adjusts pressure, giving the attached platform a steady lift. Cables from a crane lift an anchor (top left). Small items are brought up in a wicker basket (upper middle).

With a metal detector, a scientist has found a buried object. A partner fans sand toward the suction of an air lift to uncover the find.

In an air-filled "telephone booth," a diver can breathe without using scuba gear. He gives a progress report to the support ship by underwater telephone.

6

1 Voyagers of the Ancient World

On a merchant ship off the coast of Turkey, sailors reel and topple, and cargo tumbles. A wall of water sweeps toward the deck. The ship, probably driven by a storm, crashes into a mass of rocks. The vessel will sink to the floor of the Mediterranean Sea. More than 3,000 years will pass before she is seen again.

Journey To Disaster

"Metal biscuits with ears"—that's how sponge diver Mehmet Çakır described objects he saw in the sea near Ulu Burun, a cape in southern Turkey. In 1983, a joint American–Turkish team went to investigate. The biscuits turned out to be large copper ingots, metal cast into easily handled shapes. These ingots had four handles—the ears. The scientists also found parts of the hull that had carried the ingots. After more study, they came to an exciting realization: The wreck would be the oldest ship ever excavated.

Cargo from at least seven lands had gone down with the ship. Before a single object was removed, the scientists mapped and photographed the site in detail for later study. The work began in 1984. By the end of four summers, the scientists had assembled a priceless collection of Late Bronze Age artifacts. The Late Bronze Age lasted more than 500 years—from roughly 1600 B.C. to 1050 B.C.

The ship's cargo originated largely in the eastern Mediterranean. That told archaeologists the vessel was westbound when disaster struck. She was carrying raw materials: ivory, glass, wood, tin, and about six tons of copper ingots—enough (when mixed with tin) to equip a small army with bronze weapons. Scientists also uncovered the oldest known folding "notebook"—an ivory-hinged tablet that once had waxed writing surfaces. The artifacts helped archaeologists date the sinking at roughly 1,350 years before the birth of Christ.

For unknown merchants in the Bronze Age, the sinking represented a financial disaster. Their loss, however, is our gain. Eventually, experts studying the wreck will piece together its story, adding richly to our knowledge. "The wreck off Ulu Burun," says archaeologist Cemal Pulak, "is a slice of history frozen in time."

MERCHANTS' WEIGHTS, USED FOR PRICING GOODS

EASY DOES IT! Off Ulu Burun, in Turkey, divers use an air-filled balloon and a tray to raise copper slab ingots—metal cast into easily handled shapes. The ship that carried the ingots went down during the 14th century B.C.

LET'S MAKE A DEAL. An ancient coastal trader unloads goods in an Egyptian port (below). Porters, top row, carry ostrich eggs, spears, and pottery, as well as copper ingots for making bronze weapons. Blue glass ingots (in baskets, bottom row) and ivory are also unloaded. Clay jars contain a variety of materials, including fragrant substances possibly used in making perfumes. Egyptian merchants weigh goods— including hippopotomus teeth, for carving—and bargain with robed Syrian merchants. On a tablet, an Egyptian clerk keeps records. This painting, done in the style of ancient Egypt, is based on artifacts found at the Ulu Burun site.

One part tin and nine parts copper make bronze—a hard, long-lasting metal. Here, a metalsmith splits a bun-shaped tin ingot into small pieces.

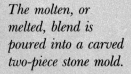

The smith melts the metals together over a fire in a vessel called a crucible.

The molten, or melted, blend is poured into a carved two-piece stone mold.

After the metal cools, a bronze sword is removed. The metalsmith inlays the handle with wood and ivory. Once sharpened, the sword is ready for action.

STAB IN THE DARK. In the depths off Ulu Burun, an archaeologist finds a real prize: a bronze sword (below). It is, experts say, of a type carried by men of wealth. Perhaps the captain of the Ulu Burun ship used it to fight off pirates. Or the weapon may have belonged to a traveling merchant. The ship sank late in the Bronze Age, which lasted nearly 2,000 years—from about 3000 to 1050 B.C. During that period, most tools and weapons were made of bronze (box, left). Iron, a much harder metal, came into general use later.

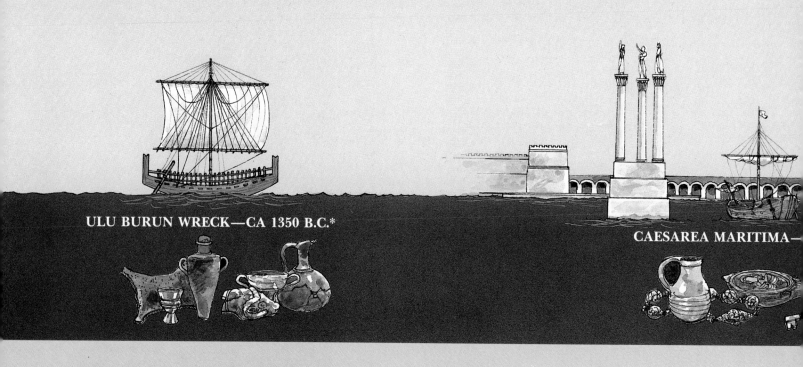

ULU BURUN WRECK—CA 1350 B.C.*

CAESAREA MARITIMA—

Western Culture's Distant Beginnings

Civilization begins in the Middle East and moves westward. The Greeks contribute ideas such as democracy, creating the foundations of Western culture. Rome adds its customs and spreads Western culture throughout the Mediterranean.

IF NEWSPAPERS HAD EXISTED in very early times, they surely would have run this headline some 10,000 years ago: CIVILIZATION IS BORN AS FARMING ATTEMPTS SUCCEED. That story actually developed over centuries in Southwest Asia and northeastern Africa, the area now called the Middle East. Our story of sea treasures begins thousands of years later, with a shipwreck off Ulu Burun, in Turkey.

By the time of the Ulu Burun wreck, about 1350 B.C., the eastern Mediterranean Sea had become a crossroads for trade. One of the trading peoples—the Greeks, or their ancestors—often settled near the coast, where they could easily trade goods and ideas. They became artists, writers, scientists, and thinkers—and they created the foundations of Western culture. In the Greek city-state of Athens, a new idea—democratic government—took root about 500 B.C. Its promise would affect the entire world.

Meanwhile, to the west, a new power was rising. Rome was a small, weak city when it declared itself a republic in 509 B.C. Some 500 years later, at the time Christ was born, Rome stood at the center of an empire that stretched from Spain to Egypt. Rome added to Greek culture such things as systems of law and a common language. The new culture spread throughout the Roman Empire.

The great port of Caesarea Maritima (pages 18–21) was built in the golden days of the empire. But empires fall, and Rome was no exception. The empire became unmanageable. Citizens strained under heavy taxes. Sometimes the ruling class acted irresponsibly. The emperor Caligula (pages 22–23) gives an early and extreme example of such behavior.

Invasions by barbarians—people outside the Roman culture—weakened Rome. The last emperor there was overthrown in A.D. 476. The event marked the end of the ancient age and the beginning of the Middle Ages.

CA 10 B.C.

LAKE NEMI BARGES—CA A.D. 40

*"Ca" indicates "about."

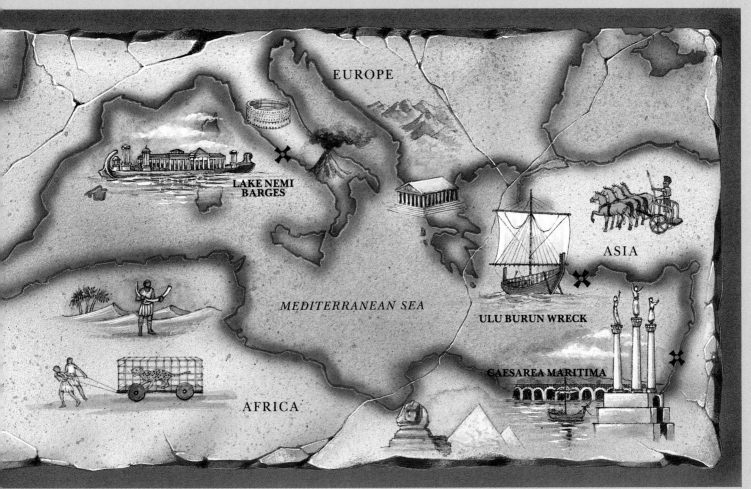

EUROPE

LAKE NEMI
BARGES

MEDITERRANEAN SEA

ASIA

ULU BURUN WRECK

CAESAREA MARITIMA

AFRICA

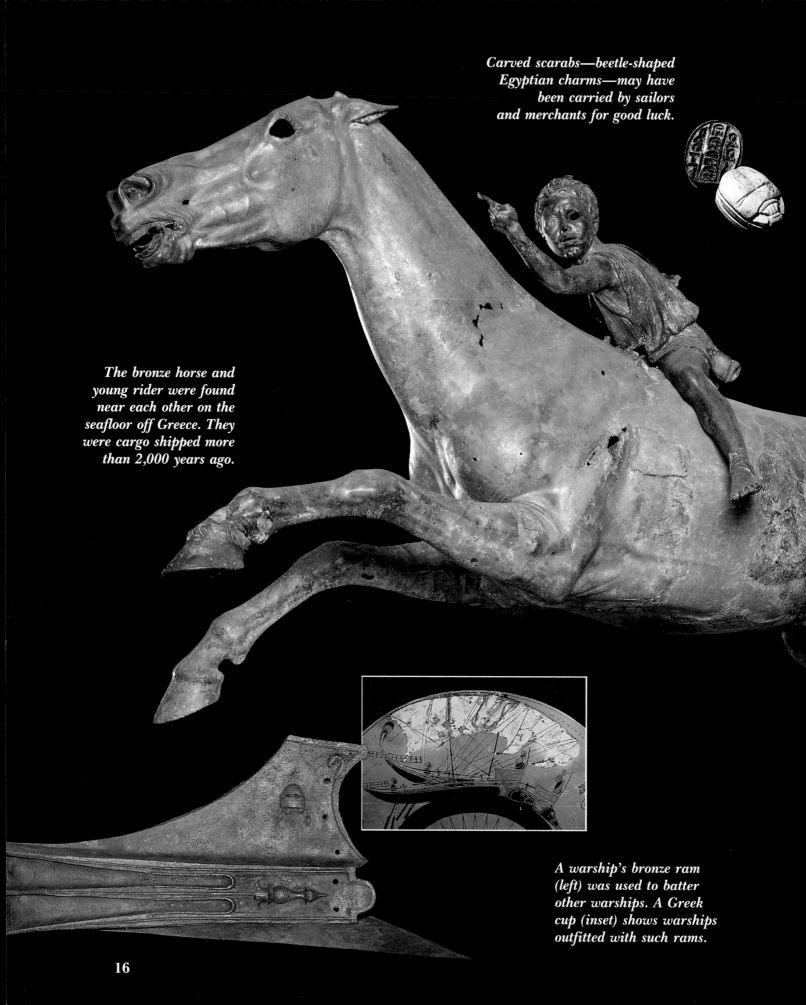

Carved scarabs—beetle-shaped Egyptian charms—may have been carried by sailors and merchants for good luck.

The bronze horse and young rider were found near each other on the seafloor off Greece. They were cargo shipped more than 2,000 years ago.

A warship's bronze ram (left) was used to batter other warships. A Greek cup (inset) shows warships outfitted with such rams.

Ancient Treasures

The sea can separate countries. It can also join them. It can be a barrier—or an international highway upon which goods and ideas are traded. For thousands of years, people of the Mediterranean have used the sea in this way. In ancient times, as now, coastal traders sailed from port to port. Sometimes the ships went down. The objects on these pages were recovered from the seafloor after being lost for many centuries.

Almonds—nearly 10,000 of them—turned up in the hull of a 2,300-year-old ship found off the coast of Cyprus. The bag they spill from is new.

Ancient birdbath? No, the Greeks used these basins (right and below) for bathing and in religious ceremonies.

17

Impossible Seaport

On the one hand, the site was no place to build a major seaport. It lacked any form of natural shelter. It stood exposed to the open sea, which ate away at the shoreline. Strong currents choked its waters with sand.

On the other hand, the site lay at a crossroads. Here, at the western edge of Judaea (now a part of Israel), inland trade routes met the Mediterranean Sea. A port meant income. Just as important, it would give King Herod I, ruler of Judaea, a chance to impress his patron, or master. The patron was Caesar Augustus, Emperor of Rome. Construction began in 22 B.C. Flattering the emperor, Herod named the port Caesarea Maritima.

Herod's engineers faced new problems that required new solutions. First: How to create a harbor on a straight coastline? The engineers came up with two massive breakwaters—arms that formed a harbor

TEAMS OF LABORERS build a breakwater, or seawall, for the port of Caesarea Maritima (right), in what is now Israel. Workers, right, tow a wooden form into place. A concrete mix is lowered into another wooden form, left. The concrete will harden underwater, forming huge blocks over a foundation of boulders already laid. More boulders, far left, are dropped to add strength. Stone slabs top off the breakwater. The port was completed about 10 B.C. Some 2,000 years later, a diver (above) recovers an ancient pot from the site, parts of which now lie underwater.

18

area and blocked the surf. The engineers built into the breakwaters a system for collecting sand-free water from wave tops and releasing it periodically. The system flushed out sand buildup. The new city had no fresh water for drinking or for watering crops. Thousands of workers built a nine-mile-long water channel, called an aqueduct, from mountain springs to the city. On the way, they had to tunnel through four miles of rock.

Caesarea Maritima served for centuries as a busy seaport and as a center of culture. But Herod had only battled nature, not conquered it. Caesarea was built on a weak spot in the earth's surface. Parts of the port gradually slipped into the sea. Today they lie 20 feet beneath the surface. Still, archaeologists marvel that the port could have been built at all. Engineers designing ports today use the same technology devised by Herod's engineers some 2,000 years ago.

ROMAN COINS (LEFT AND BELOW); A TOKEN (BOTTOM) SHOWS SHIPS APPROACHING CAESAREA MARITIMA

BUSY SEAPORT. The most skilled engineers, recruited from Rome, designed Caesarea. Thousands of laborers built it. Ships lowered their sails before entering the harbor. Rowboats may have then towed them in (above). In this reconstruction, a lighthouse stands at the end of one breakwater; the harbormaster's building stands at the end of the other. As many as a hundred ships could anchor in the harbor.

Herod I, King of Judaea, ordered Caesarea built for the taxes it would generate and to gain favor with Caesar Augustus, Emperor of Rome. His port city would have parks, statues, office buildings, a theater, a huge library, a circus, luxurious public baths, and a stadium that could seat 38,000 people.

An aerial view (right) shows what remains of Herod's port facility. It now lies underwater. Here, a white diving barge lies in the channel between submerged breakwaters.

20

LAMP FOUND AT
CAESAREA BURNED
OLIVE OIL

The Emperor's Pleasure Barges

He was definitely strange and perhaps insane. His cruelty made him a legend. Sometimes, an ancient biographer reports, he would order criminals fed to wild animals in the sports arena. He claimed to be a god—and dressed the part in flowing gowns. He would swallow expensive pearls, being both a wild spender and a show-off. It was rumored he nominated his horse for consul, a high governmental post. He was Caligula, Emperor of Rome from A.D. 37 to 41.

Insane or not, Caligula clearly liked high living. On becoming emperor, he ordered the construction of two floating palaces. The vessels were probably built at public expense. Whether Caligula ever invited the public aboard is not known. It is believed, however, that the barges served mainly for his personal pleasure.

The vessels held splendid courtyards and magnificent slumber and banquet rooms (right). They had luxurious baths with running water and employed an army of servants. Nothing quite like the pleasure barges has been seen since Caligula's day.

In A.D. 41, Caligula, 28, was assassinated. His barges, at some unknown time, sank. For centuries, people living near Lake Nemi, outside Rome, knew that ancient vessels lay on the lake bottom. Very likely they were Caligula's barges. Not until 1928, however, did anyone make an all-out attempt to raise the ships. In that year, Italy's leader, Benito Mussolini, ordered Lake Nemi drained. The job took several years, but finally the hulls became visible. They were restored and placed on display in a lakeside museum.

The display, however, was to be short-lived. In 1944, during World War II, fire destroyed the hulls. Now the barges exist only in drawings and old photographs—and in the imagination.

DECORATIVE BRONZE
WOLF HEAD

BARE BONES. Workers stand beside the exposed hull of a 2,000-year-old barge, one of two salvaged in Lake Nemi, in Italy. The barges may have been the ones built by the Roman emperor Caligula. Though probably paid for by Roman citizens, the barges served as Caligula's own pleasure palaces. This rare photograph gives barely a hint of the barges' former splendor. At some point, the Lake Nemi barges sank—no one knows when or why. Only in the 1930s were the hulls recovered. They were restored and put on display—but not for long. During World War II, the hulls were destroyed by fire.

LUXURY, ROMAN STYLE. Caligula had a taste for high living. This painting shows the emperor (in dark purple tunic, or gown) entertaining aboard one of his pleasure barges. Lounging on couches, visitors enjoy a meal in a richly furnished banquet room. The food, of course, is the best that Rome can provide. Out of sight, servants stand ready to jump at the snap of a finger. In the distance lies the second barge.

After the meal, the guests can stroll through any of several courtyards.

The courtyards have as centerpieces elegant fountains and statues. If the breeze from across the lake becomes chilly, guests may duck under a covered walkway and pick fruit from trees that grow there. Running water is piped throughout the barge. Guests can, if they wish, take a relaxing bath in their choice of pools before turning in for the night in a luxurious bedroom.

This scene is based on writings of the time, on artifacts found at Lake Nemi, and on knowledge of Roman architecture. Experts say it probably gives a reasonably accurate idea of the high life aboard Caligula's pleasure barges.

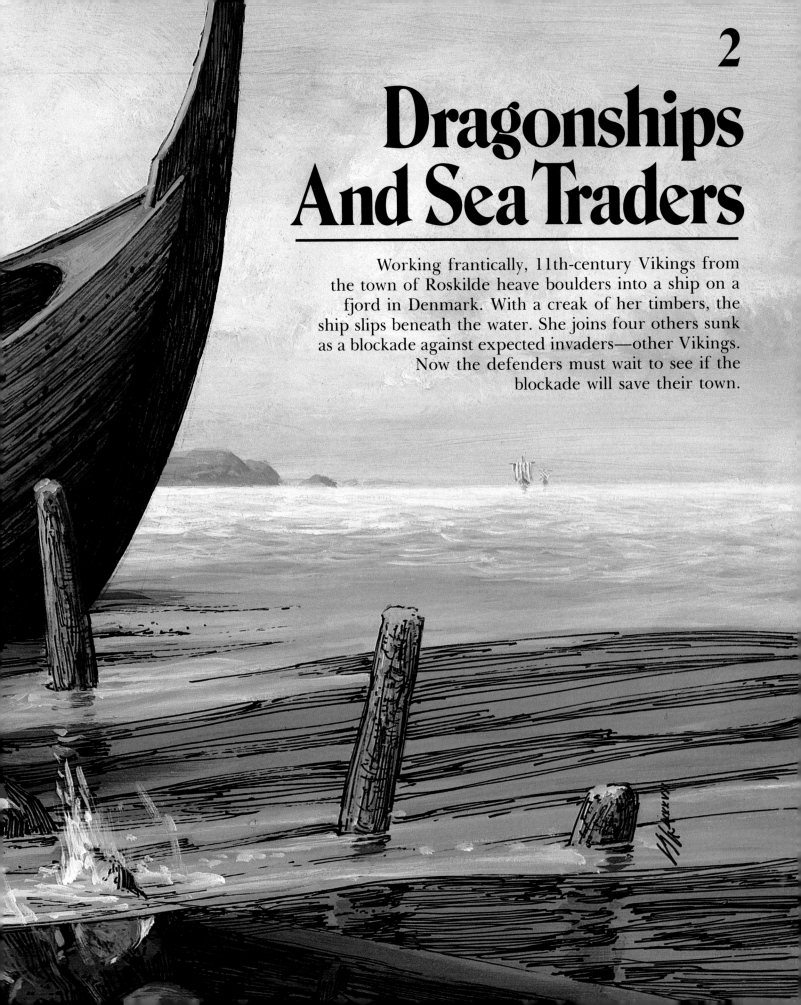

Dragonships And Sea Traders

Working frantically, 11th-century Vikings from the town of Roskilde heave boulders into a ship on a fjord in Denmark. With a creak of her timbers, the ship slips beneath the water. She joins four others sunk as a blockade against expected invaders—other Vikings. Now the defenders must wait to see if the blockade will save their town.

Bold Rovers, Sturdy Ships

For some 250 years they terrorized much of Europe. Appearing in their fearful dragonships, they raided and colonized areas from the western Atlantic Ocean to the Caspian Sea. They were the Vikings—Danes, Swedes, and Norwegians who roved the seas as pirates and traders.

Master mariners, Vikings explored uncharted lands. They were probably the first Europeans to reach the North American mainland. They discovered Greenland. They built a bustling network of trade routes, and their goods reached as far as China.

The Viking age lasted from about A.D. 800 to 1050. Although ships played an important role in that period, not much was known about working Viking craft until fairly recently. A few ships had been recovered, but they were built as royal barges. Then, in 1957, an important discovery was made at Skuldelev, on the Roskilde Fjord.

Local people had long known that *something* was at the bottom of the fjord (fee-ORD, a narrow inlet of the sea). For centuries fishermen had complained that their boats scraped bottom in a certain area. Finally, divers went down for a look. They came up with startling news. Five Viking ships lay at the bottom of the fjord!

To recover the ships, engineers first built a watertight wall called a cofferdam around the wrecks. Then they pumped out the seawater and put up a walkway. Next, archaeologists used water jets to spray away layers of sand and mud and to keep the timbers from drying and

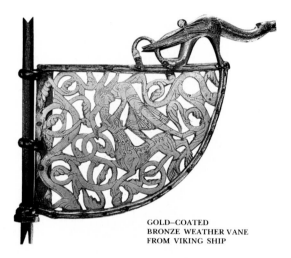

GOLD–COATED
BRONZE WEATHER VANE
FROM VIKING SHIP

shrinking. The scientists then lifted out the rocks that had sunk the ships. Finally, the timbers were removed for preservation.

Two of the craft were swift, sleek dragonships, or longships, used to wage war. The third ship was a merchantman—a trading vessel—called a *knarr* (kuh-NAR). Its high sides protected crew and cargo on the open sea. The fourth ship was a coastal trader that hauled goods on fjords and on the Baltic Sea. The fifth was probably a ferry or a fishing vessel.

Now in a museum at Roskilde, the ships demonstrate the Vikings' seafaring skills. But did their deliberate sinking spare the town? That's a mystery the ships can never solve.

PREPARING FOR CONQUEST, woodsmen (below, at left) chop down trees. The trees will become the ships that will carry the soldiers who will conquer England. This scene is part of a 231-foot-long embroidery called the Bayeux Tapestry. It's an important source of information on customs of the time. The work shows how Viking descendants conquered England in 1066. They crossed the English Channel from France in about 800 ships similar to the Roskilde villagers' ships—found in 1957 near Skuldelev, Denmark. Here, craftsmen, center, form the timbers and planks. Workmen, right, drag finished ships down to the water.

HIC TRAHVNT:NAVES:ADMARE:

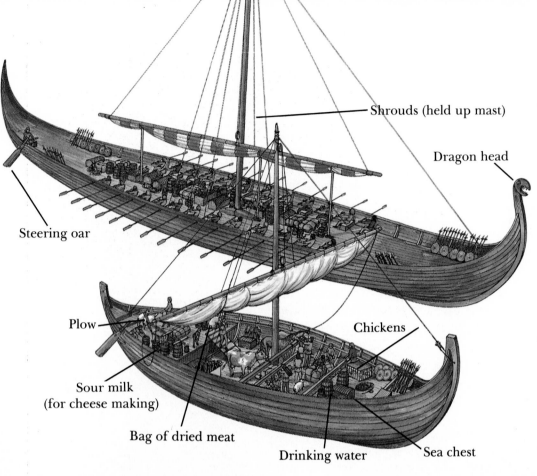

Shrouds (held up mast)

Dragon head

Steering oar

Plow

Sour milk
(for cheese making)

Chickens

Bag of dried meat

Drinking water

Sea chest

BUILT FOR THE SEA. In addition to carrying 40 to 60 warriors, the longship (left, top) transported weapons and supplies. These long, narrow craft cut swiftly through the water under sail or oar power. Fierce figureheads earned them the name "dragonships." For long voyages with heavy cargoes, the Vikings relied on the shorter, broader knarr *(kuh-NAR). These deep-sea traders (left) carried families and merchants to distant lands. Viking shipbuilders fixed the rudder to the right side of the hull. An old Norse word meaning "steering side" has come into English to mean "starboard"— a ship's right side.*

RECOVERING TIMBERS, archaeologists (left) remove rocks that were used to sink the Skuldelev ships nearly a thousand years ago. They work within a watertight steel enclosure called a cofferdam. It holds back the fjord, permitting the scientists to excavate the ships. A constant spray of water from sprinklers helps preserve the timbers—and makes rain gear required dress. The vessels now stand in a museum at Roskilde. This scene (above) shows a knarr, in front, and a coastal trader, left. Danish Boy Scouts built a copy of a Skuldelev longship (right). The ship, they report, gives a smooth, fast ride.*

SKULDELEV SHIPS—CA 1000

GLASS WRECK—CA 1025

Shifting Fortunes in the Middle Ages

As one empire falls in Europe, another rises in the Middle East.

WITH THE FALL OF ROME to barbarian invaders, the Middle Ages began. That period lasted roughly a thousand years, ending about 1500 with the dawn of modern civilization.

You might wonder: If barbarians captured the Roman Empire, why wasn't Western culture lost? The answer lies in the church. Before Rome's fall, Christianity had become the official religion. The church had strong leaders throughout the empire. They preserved much of the culture the Romans had built and added to it the teachings of Christianity.

The invaders added customs from their own cultures. The Vikings, for example, were keen traders. By the time of the Roskilde sinkings, they had set up thriving trade routes throughout Europe.

In the Middle East, a new culture was rising: the Islamic, or Muslim, Empire. At its core was Islam, a religion founded in the 600s by an Arabian merchant named Muhammad.

Islam had huge appeal; it spread explosively. By 732, the Muslim Empire stretched from southern Spain to the frontiers of China. Muslims prized knowledge, and the empire became a leader in learning and culture. Among countless other contributions, the Muslims introduced Europe to the numerals 0 through 9. (Imagine doing multiplication with the old system—Roman numerals!)

Muslim merchant ships— including, perhaps, the glass wreck (pages 32–35)—dominated trade with the Far East for hundreds of years. They went as far as China. (That ancient culture has come up with many important inventions. One in particular, developed at least 300 years before the Sinan wreck [pages 36–37], would change history. It was gunpowder.)

War and internal conflict eventually broke up the Muslim Empire. In Europe, kingdoms began to rise from the barbarian states. Those kingdoms would, for the most part, become the nations you see on a map of Europe today.

SINAN WRECK—CA 1330

ATLANTIC OCEAN

KULDELEV
SHIPS

EUROPE

MEDITERRANEAN SEA

AFRICA

ASIA

GLASS WRECK

SINAN
WRECK

PACIFIC
OCEAN

Puzzle of the Glass Wreck

The diver plunged into the waters of Serçe Limanı, a harbor on Turkey's southwest coast. He filled his bag with sponges and prepared to surface. Just then, he glimpsed colorful pieces of glass in the sand. He grabbed some and swam up. That might have been the end of the story if a scientist named George Bass hadn't come on the scene.

Dr. Bass, a nautical archaeologist, was working in Turkey when he heard about the find. He did further research, developing a strong hunch that an 11th-century wreck lay in the harbor. In 1977, Dr. Bass led a team of Turkish and American archaeologists to Serçe Limanı.

After only the second dive, the scientists began to uncover a wooden hull resting in the sand at 110 feet. Thousands of glass fragments lay on and around it. The team found coins and glass weights that dated the sinking at about A.D. 1025. Dr. Bass's hunch had paid off well. The glass wreck was a rare find: an early Mediterranean example of "modern" ship construction (see page 34).

Artifacts from the wreck present a riddle. Amphoras (right) bear lettering of the Christian world. Other objects bear images of the Virgin Mary and of other Christian figures. Elegant glassware (left) found intact—ready for use—was crafted by Muslims, as were the glass weights. But Christians and Muslims were bitter enemies at the time. Where did the ship come from? How did she happen to be carrying goods from both worlds? It may take years to learn the answers.

In addition to the undamaged items, the archaeologists found tons of broken and raw glass. Did a shipment of fine glassware become shattered en route? "Not at all," says Dr. Bass. "The scrap was simply part of the cargo. The ship was hauling it for remelting—a centuries-old example of recycling!"

TUMBLER AND BOTTLE FROM GLASS WRECK

GLASSMAKER'S ART. Muslim craftsmen (left) make glassware such as that recovered from an 11th-century shipwreck off Turkey. A boy, left, feeds a fire. His father, center, blows a red-hot glob of molten glass into a bottle shape. An assistant, right, cracks a bottle off a blowpipe. Broken glass, in front, will be melted down for reuse.

KNEELING AMONG AMPHORAS, a diver (right) inspects an ancient bottle found at the stern of the glass wreck. Archaeologists began excavating the wreck in 1977. Glass pieces brought up by a sponge diver had touched off the search.

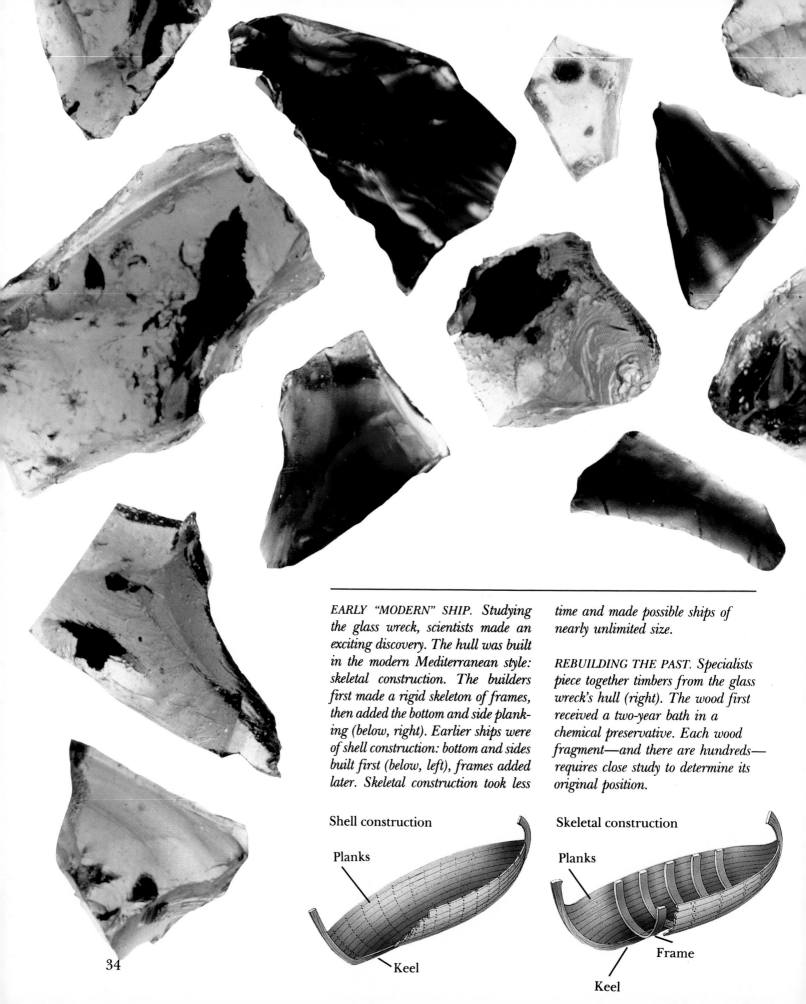

EARLY "MODERN" SHIP. *Studying the glass wreck, scientists made an exciting discovery. The hull was built in the modern Mediterranean style: skeletal construction. The builders first made a rigid skeleton of frames, then added the bottom and side planking (below, right). Earlier ships were of shell construction: bottom and sides built first (below, left), frames added later. Skeletal construction took less* time and made possible ships of nearly unlimited size.

REBUILDING THE PAST. *Specialists piece together timbers from the glass wreck's hull (right). The wood first received a two-year bath in a chemical preservative. Each wood fragment—and there are hundreds—requires close study to determine its original position.*

Shell construction

Planks

Keel

Skeletal construction

Planks

Frame

Keel

34

RAW GLASS—about two tons' worth, in chunks—made up most of the wreck's cargo (left). The ship was hauling the scrap for recycling when she sank in the harbor of Serçe Limanı about 960 years ago.

PICKING UP THE PIECES. The ship also carried broken finished glassware for recycling. About a million fragments of it have been found. Here a glass expert pieces together fragments. The wreck is a giant jigsaw puzzle that will take many years to solve.

Hard Pillow At Sinan

Waves crashed across the ship's deck. Wind and rain whipped her sails. Fearing disaster on the open Yellow Sea, the captain steered toward a cluster of islands in Sinan, an area on the Korean coast. As the crew fought to bring the ship between two rocky islands, whirlpools churned the water. Perhaps spinning out of control, the ship disappeared into the sea. There, as far as we know, she lay unnoticed for some 650 years.

In May 1975, a fisherman hauled in his net from the waters off Sinan. Instead of fish he found six vases encrusted with minerals. Experts identified the bluish green porcelains, called celadons, as objects from 14th-century China.

The next year, a team of South Korean Navy divers explored the site. The water was so cold that some divers bit off their scuba mouthpieces in involuntary reaction to it. Swift currents stirred up the muddy bottom, making it impossible to see. Working by touch alone, the divers slowly uncovered the ship's remains.

The ship was apparently a merchantman hailing from China. She was carrying goods, probably to a port in Japan, when the storm hit. Her holds bulged with a varied cargo. The bulk of it consisted of celadon ware—some 600 bowls (left), platters, vases, statuettes, and incense burners. There were also bronze mirrors, pepper, tons of Chinese coins, and even a porcelain pillow. (Sleepers enjoyed the cool, if hard, surface.)

Many of the items lay neatly packed in wooden crates. The divers recovered ten such crates from the treacherous waters where the ship and her crew had perished. Marked on one were the Chinese characters 大吉 —"great luck."

CLAWS OUTSTRETCHED, a dragon struts across a 600-year-old Chinese bowl (above). In 1976, divers recovered the bowl and thousands of other ceramic pieces from a wreck off Sinan, in South Korea. Some Chinese believed that this blue-green porcelain, called celadon, would break or change color if poisoned food touched it. It didn't, of course. Still, the belief no doubt helped merchants sell a lot of celadon ware.

FACE-LIFT. Technicians use acid to remove marine growth from pots and jars found in the wreck (right). Gloves and face masks protect the men from the skin-eating chemicals. A freshwater bath (far right) draws harmful salts out of celadon and other pottery. Treasures from the Sinan wreck had lain in salt water since about 1330. They require careful restoration to prevent cracking and fading.

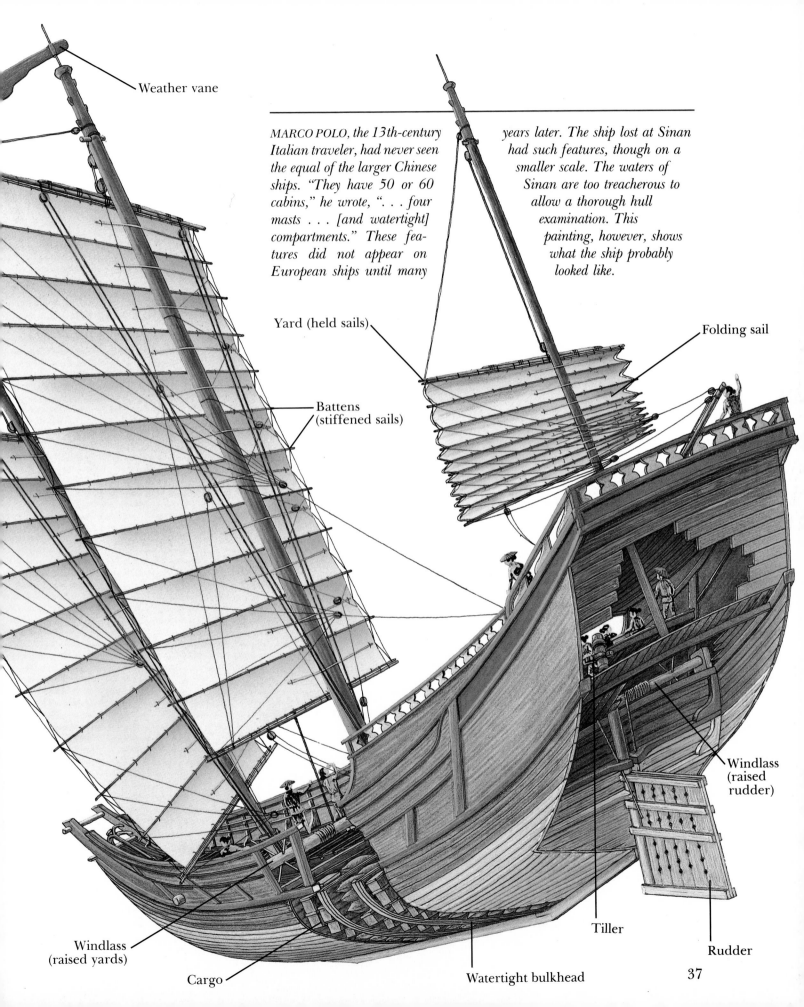

Weather vane

MARCO POLO, the 13th-century Italian traveler, had never seen the equal of the larger Chinese ships. "They have 50 or 60 cabins," he wrote, ". . . four masts . . . [and watertight] compartments." These features did not appear on European ships until many years later. The ship lost at Sinan had such features, though on a smaller scale. The waters of Sinan are too treacherous to allow a thorough hull examination. This painting, however, shows what the ship probably looked like.

Yard (held sails)

Folding sail

Battens (stiffened sails)

Windlass (raised rudder)

Windlass (raised yards)

Tiller

Rudder

Cargo

Watertight bulkhead

37

3
Whalers and Men-of-War

With one lash of its powerful tail, a harpooned whale upends a boat, pitching sailors into the sea. The time is the 1500s, and the men have sailed some 3,000 miles to hunt whales in North American waters. Oil from the whales will bring a high price in Europe—rewarding the whaling men for their hard, dangerous work.

Bad Wind At Red Bay

You might call it the world's first oil boom. Every spring and early summer from about 1540 to 1600, dozens of ships set off on a westerly course across the Atlantic. The ships were manned by Basques, people from the border region of France and Spain. Their destination: Labrador, a large peninsula on the east coast of Canada.

The voyage lasted about a month. The men stayed in Labrador for as long as six months, hunting whales for their oil. The oil lit lamps throughout Europe. It was a major ingredient in some soaps and medicines. A large shipload of whale oil might sell in Europe for today's equivalent of six million dollars.

In 1565, in the large harbor of Red Bay, Labrador, a ship lay at anchor. She was fully loaded for the voyage back to Europe. But a violent autumn storm swept down from the Arctic. It ripped the ship from her moorings and sent her to the bottom.

For 413 years, the ship lay hidden. A modern-day researcher, Selma Huxley Barkham, suspected that Red Bay might hold such a wreck. Going by her lead, archaeologists from the Canadian Park Service discovered timbers at the bottom of the bay in 1978. Old records helped them identify the wreck—though without certainty—as the

BASQUE WHALEMAN'S
IRON HARPOON HEAD

OIL FOR EUROPE'S LAMPS. In this old woodcut, workers remove whale blubber. Melted down, the blubber will yield an oil widely used in Europe. In the 1500s, men from the Basque region of France and Spain made yearly voyages to Labrador, in Canada, to hunt whales for their oil.

IN THE ICY WATERS of Red Bay, in Labrador, archaeologists use a grid to map 400-year-old timbers from a wreck (right). It's probably the Basque whaler San Juan. *The ship sank at anchor during a violent storm in 1565.*

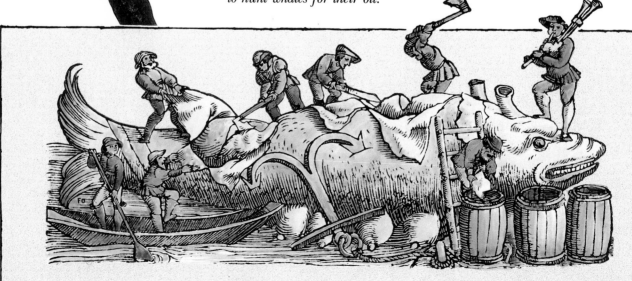

Basque whaler *San Juan*. It would prove to be the earliest largely intact ship yet excavated in the Americas.

The icy waters made excavation difficult. Divers wore heated wetsuits. Dangerous icebergs passed nearby. Cold-water fish called sculpins kept trying to snatch the scientists' pencils!

But the cold also had helped to preserve the ship. A whaleboat in good condition was pinned under the ship's hull. Below the collapsed decks lay the remains of 450 huge oak barrels that once held whale oil. Large numbers of codfish bones gave clues to the crew's diet. Apparently all hands survived the sinking, for no human remains were found. But next to the codfish remains, archaeologists did find some puzzling bones. They turned out to belong to the only known casualty—a black rat.

SANDS OF TIME. Sandglasses like this one found at the **San Juan** *site (above) helped navigators calculate a ship's speed. A wooden float was tied to a line and dropped off the stern. Sailors (top) let the line pay out while a cabin boy prepares to call out "Mark!" at 30 seconds as measured by the sandglass. Relating time to the amount of line paid out gave ship's speed. Some time later, sailors began tying knots in the line at regular intervals to speed the measuring. Ever since, sailors have expressed ship's speeds in knots, as: "The ship is making 24 knots"— 24 nautical miles an hour. That's 27½ land miles an hour.*

MARY ROSE—1545

SAN JUAN—1565

Great Age Of Discovery

The need to open up new trade routes to the Orient touches off the great age of discovery. An especially productive part of that period lasts about 30 years. In that time, the known world roughly doubles in size.

MARCO POLO, the 13th-century Italian traveler, had sparked Europe's interest in the Orient. Polo had brought back stories of fabulous riches to be found in Asian lands. But Muslim merchants controlled the known routes to those lands—and were not eager to share them.

At the time, shipbuilders were designing larger, stronger hulls and more efficient sail combinations. The new designs suited mariners sailing open seas on long voyages of discovery. Of those voyages, three especially stand out.

● In 1492, the Italian Christopher Columbus sailed into the Caribbean, sighted land, and "discovered" America for Spain. (The Viking landings were long forgotten.) Discovering a continent was almost certainly the furthest thing from his mind. Columbus, at first, thought he had found a short route to Asia.

(A German mapmaker misnamed the new land America. He believed an explorer named Amerigo Vespucci had discovered it. By the time he realized his mistake, the name was firmly rooted.)

● In 1497–98, the Portuguese navigator Vasco da Gama sailed around the tip of Africa and reached India by sea. That voyage opened a new route to the Orient. Merchants would rush to use it.

● In 1519–21, the Portuguese Ferdinand Magellan completed Columbus's dream. He found the westward route to Asia.

Only a few dozen years after Columbus's landing, an era of worldwide trade began. Hundreds of merchant ships such as *San Juan* sailed the open seas, laden with sugar and whale oil as well as silver, silk, and spices.

As shipping lanes gained importance, rulers built powerful craft to defend them. *Mary Rose* (pages 44–49) and *Vasa* (pages 50–55) were designed for that purpose. Today's big warships trace their ancestry to such mighty men-of-war.

VASA—1628

NORTH
AMERICA

SAN JUAN

ATLANTIC
OCEAN

MARY ROSE

VASA

EUROPE

AFRICA

The Pride Of King Henry

MEETING THE KING. History fan Elizabeth Iaukea, 15, who lives in Tallahassee, Florida, studies a wax figure of England's King Henry VIII (above). In 1545, Henry watched in horror the sinking of his prize warship Mary Rose.

July 19, 1545, was a clear, calm day at the naval base in Portsmouth, England. People proudly watched as *Mary Rose*, finest ship in the royal fleet, sailed out to meet the French enemy. Suddenly, when the ship was only a mile from shore, an unexpected gust of wind caused her to heel, or tilt, dangerously. The sea rushed in through open gunports. On shore, King Henry VIII could hear the screams of drowning sailors. An onlooker fainted. In less than a minute, it was over. *Mary Rose* had sunk, taking with her almost 700 crewmen and soldiers.

Big, sleek, and heavily armed, *Mary Rose* was one of the most advanced fighting ships of the day. She filled King Henry with pride. Henry was only 18 when, in 1509, he ordered the ship built. It was later rebuilt and armed with the latest in weapons, including bronze guns weighing as much as three tons each. The extra weight made *Mary Rose* top-heavy and may have helped cause her to capsize.

For more than 400 years, *Mary Rose* lay on her side at the bottom of the English Channel. Layers of silt protected her. When archaeologists began excavating the ship, they found things much the same as they were on the day she sank. A surgeon's chest held a wooden jar containing ointment with finger marks still in it. Food remains—bones, plum pits, peas still in the pod—lay here and there. The skeleton of an archer carrying a bundle of arrows was found near the stairs where he had tried to escape. Other skeletons included those of a 14-year-old cabin boy and of a small dog.

In all, archaeologists removed more than 18,000 artifacts from the site. If you travel to Portsmouth, you can see many of them at the navy base where *Mary Rose* now rests. It's just a few minutes' walk from where she was built five centuries ago.

DISASTER STRUCK Mary Rose *as she sailed out of Portsmouth, England, to drive back an invading French fleet (left). A gust of wind blew her over on her starboard side. Water rushed into her open gunports, sending the ship down. Only about 30 of the 700 men aboard survived.*

BIG GUN. At a traveling exhibit in Florida, Elizabeth sights down a copy of an iron cannon from Mary Rose *(right). The ship's cannon weighed up to three tons each. Such heavy guns, placed high on board, may have unbalanced the ship and led to her sinking.*

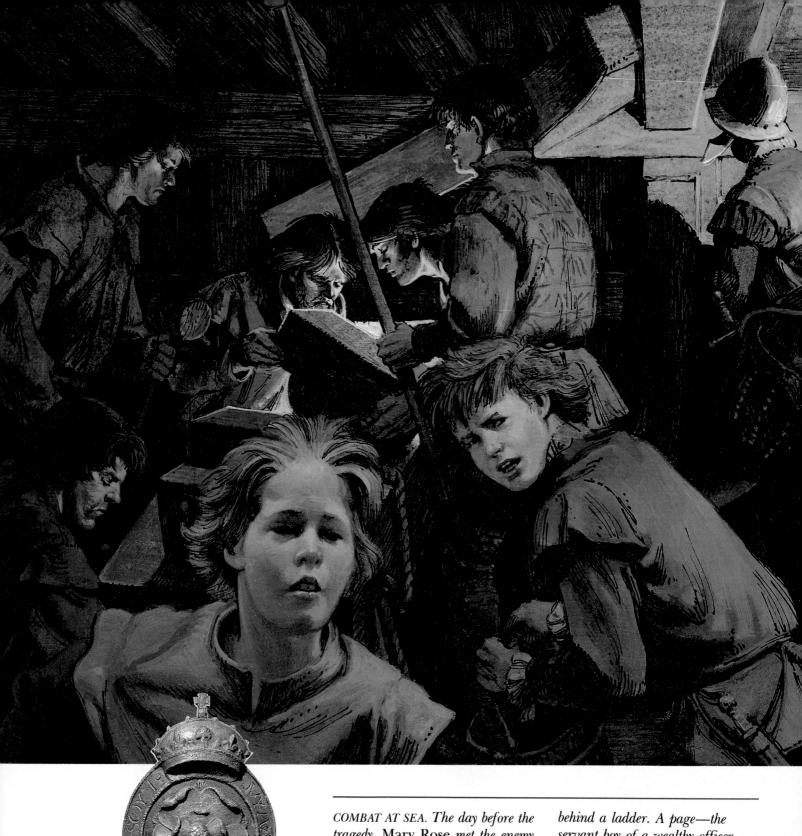

HENRY VIII'S FAMILY EMBLEM FROM A BRONZE GUN

COMBAT AT SEA. The day before the tragedy, Mary Rose *met the enemy in battle. This painting shows how the fighting may have raged. Gun crews rush to load and fire cannon aimed through open gunports. During the commotion, a barber-surgeon treats the wounded at an on-the-spot clinic behind a ladder. A page—the servant boy of a wealthy officer—dashes on some errand; another page runs with a bucket of sand to douse small fires. His bow slung over his shoulder, an archer hurries to an upper deck to join comrades raining arrows on the enemy.*

CLOSE–RANGE WEAPON
'SQUARE MURDERER'
WITH IRON HAILSHOT

ARROWS STORED IN
A LEATHER DISK

Artifacts Tell a Tale

Mary Rose yielded hundreds of everyday objects used by the crew. Researchers found clothing and sewing kits. They recovered games and musical instruments used by sailors in their few free hours. Finds such as these give us an excellent idea of life aboard a 16th-century man-of-war.

A wooden pomander (right) provided an officer with a whiff of something more pleasant than the foul air below. The pomander contained fragrant spices.

A single one of these coins (left) might have paid an officer for a day, an ordinary seaman for a month.

A wooden pocket sundial (below) helped an officer tell the time of day.

Game items found at the site include a domino and dice (below). Gambling was a favorite pastime of seamen.

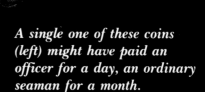

One side of this comb took care of normal grooming. The finer side was used to rid the head of lice. Another type of vermin— fleas—also pestered crewmen.

Board games helped relieve the monotony of life at sea. Two such games are carved into this barrelhead (below).

The Ship That Never Fought

To the well-wishers who watched her ease out of Stockholm harbor on August 10, 1628, the Swedish warship *Vasa* was an awesome sight. Large square sails billowed from her tall masts. Rows of bronze cannon, 64 in all, lined three upper decks. A golden lion figurehead sprang from the bow. Hundreds of carvings layered with gold decorated the ship's stern and sides.

King Gustavus II Adolphus had hired the most skillful Dutch shipbuilders to construct *Vasa*. The finest craftsmen were commissioned to decorate her. Now she was on her maiden voyage, bound for the Baltic Sea and service in a war against Poland. But less than a mile out of port, the wind heeled her over, and she capsized. About 50 people drowned. The shipbuilders and the ship's officers were summoned to testify in an investigation. The inquiry failed, however, to determine the exact cause of the disaster. Experts today know the ship was topheavy. She should not have been carrying 64 big guns so high up.

Salvage attempts undertaken in 1663–65 raised 53 cannon. Then

OFFICER'S FELT HAT WITH BAND

the ship lay largely undisturbed until 1956, her location forgotten. In that year Swedish engineer Anders Franzén succeeded in locating *Vasa* in Stockholm harbor. She was remarkably well preserved. The problem was how to raise *Vasa* without destroying her. One person suggested filling the wreck with Ping-Pong balls to float it. Another proposed freezing it, then floating it like an ice cube. You'll see on the next pages how *Vasa* actually was raised and brought ashore.

Once *Vasa* was on land, archaeologists faced the problem of preserving her. They removed tons of mud, finding 14,000 artifacts and 12 human skeletons. The ship's timbers, if kept in dry air, would have shrunk and rotted. As a temporary measure, the hull was bathed in a continuous mist. Then, the experts began spraying on a waxy chemical that penetrated each timber to the core.

The recovery and preservation effort has taken more than 20 years. Today, visitors to the Vasa Museum, in Stockholm, see the ship much as she appeared 360 years ago. *Vasa* is so well restored, say experts, that she could probably sail again—with some cannon removed.

SAILOR'S DITTY BOX, USED TO HOLD NECESSITIES

MUD OF CENTURIES blankets a gun deck of the Swedish warship Vasa *(below). The ship sank in Stockholm harbor in 1628, minutes after she began her first voyage.*

CRAMPED QUARTERS. Aboard 17th-century warships such as Vasa, *crewmen and soldiers worked and relaxed in the same area: the gun deck (left). They slept on bare planking, without mattresses or blankets. They ate mostly dry bread and dried or salted meat, with beer. When men became ill—and that was often—they were treated by the ship's barber, who doubled as surgeon. Pets, music, and games gave some relief from the often monotonous and always hard shipboard life.*

1628

1663–65

1959

REFLOATED AND RESTORED, Vasa
gained a second life after centuries
on the harbor floor. Her salvage
is regarded as an engineering feat.

1628 . . . Vasa sinks in Stockholm
harbor in 110 feet of water.

1663–65 . . . Using a simple
diving bell, men recover 53
of Vasa's cannon.

1959 . . . Divers thread cables
under Vasa's hull. The cables are
attached to pontoons—large floats—
partially filled with water. The
water is pumped out, raising the
pontoons. Slowly, Vasa rises 8 feet.
In 18 gentle lifts, Vasa's fragile
hull is "walked" 1,969 feet up the
sloping harbor floor. It comes to
rest in 50 feet of water. Divers
prepare the hull for another lifting.

1961 . . . Vasa is raised again
and floated into dry dock.
Archaeologists remove mud and
artifacts. Finally, the ship is towed
to a new site for preservation.

1961

AFTER BEING RAISED, the blackened Vasa rests in a partially filled dry dock (below, left). Long poles help prop up the ship, which still holds tons of mud. A continuous spray of water keeps the timbers moist to prevent shrinkage and rot. Preservation of Vasa's hull continued indoors with chemical treatments. By 1980, 21 years after the raising, restoration was nearly complete (below). Today, the ship rests as a national treasure at Stockholm's Vasa Museum.

CARVED WOODEN
FIGURES
FROM *VASA*

FEARSOME FACES like the one at left were intended to guard Vasa in battle. The carvings were attached to the insides of the gunport lids. When gun crews swung the lids open, enemy soldiers would have faced two rows of fierce, snarling "lions."

ALL HANDS stand by their work stations as Vasa gets under way (right). Large for her time, Vasa had four decks and a hold. The ship would have taken up two-thirds the length of a football field. From bottom to top, the stern rose as high as a six-story building. With her rows of elaborate gold-layered carvings, the ship was designed to impress the enemy with Sweden's power.

Captain and
first mate

Tiller (moved rudder)

Rudder

Military supplies

Bilge
(held dirty water)

Whipstaff
(moved tiller)

Bilge pump

Upper deck

Gun deck

Lower gun deck

Capstan
(lifted heavy
weights)

Orlop deck

Hold
(carried cargo)

Rock ballast
(steadied ship)

Food and
beer stocks

Galley

Sail locker

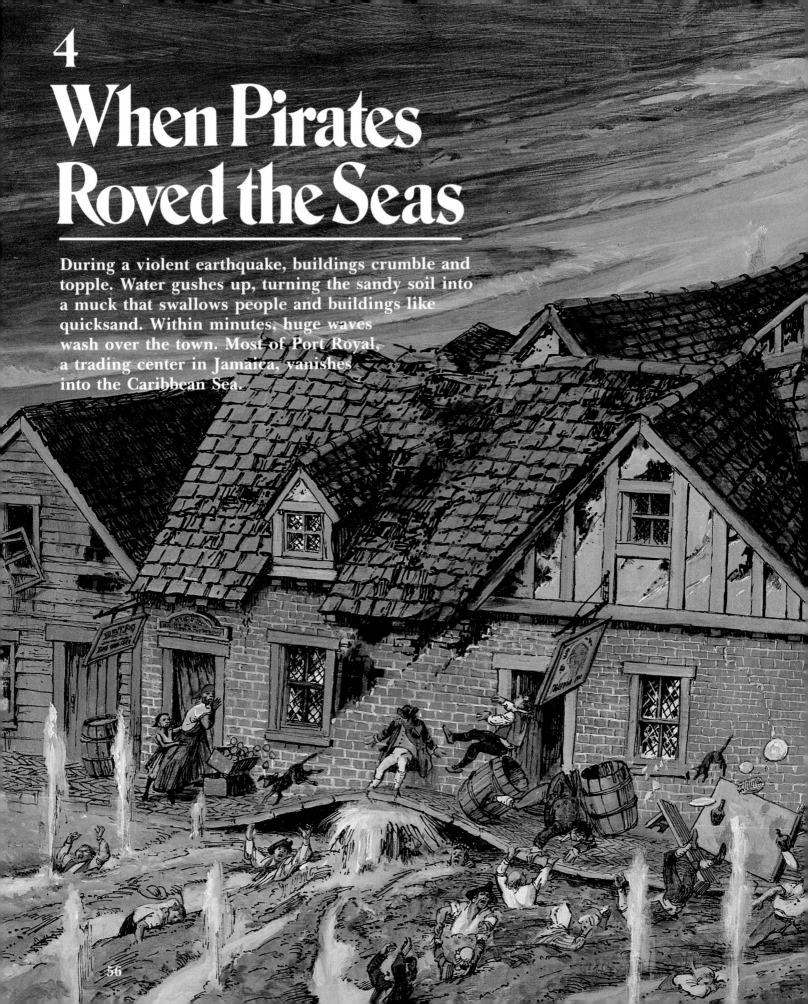

4 When Pirates Roved the Seas

During a violent earthquake, buildings crumble and topple. Water gushes up, turning the sandy soil into a muck that swallows people and buildings like quicksand. Within minutes, huge waves wash over the town. Most of Port Royal, a trading center in Jamaica, vanishes into the Caribbean Sea.

56

'World's Wickedest City'

A brass pocket watch recovered from the site tells part of the story. It stopped at 17 minutes before noon—the time of the earthquake that destroyed Port Royal, on the island of Jamaica. Two-thirds of the town was plunged underwater. More than 2,000 people perished instantly; within weeks, some 2,000 more had died of injury and illness.

Port Royal had a reputation as the "wickedest city on earth." For years it served as a pirate haven. The English had captured Jamaica from the Spanish in 1655. They invited pirates to use Port Royal as a safe place to sell their stolen goods. In return, the pirates raided ships and settlements of England's chief rival, Spain. Later, however, relations between the two nations improved. Laws against piracy were enforced, and Port Royal became a major trading center.

By the time of the earthquake, in 1692, Port Royal was one of the richest English towns in the New World. More than 6,000 people lived and worked in its 2,000 buildings. Markets, warehouses, taverns, gambling dens, churches, and houses lined the narrow streets.

Archaeologists, with the support of the Jamaican government, are now conducting a detailed underwater excavation of Port Royal. They're focusing at present on the town's main business area. The scientists have found a six-room building with brick floors (right). In one room, they uncovered a mound of 60 onion-shaped bottles. That room, they say, was probably used as a tavern. Another room, littered with scraps of leather and wooden objects, may have housed a shoemaker and a woodworker.

A lifetime or more will pass before scientists complete the study. Already, though, the work is giving us a fascinating view of life in the rowdy, buccaneering old town of Port Royal.

PEWTER TANKARD
FOUND IN PORT
ROYAL EXCAVATION

BRASS
CANDLESTICK

DISASTER AREA. The white line (left) shows the outline of Port Royal before the earthquake of 1692. Much of the area that plunged into the sea has since been filled in. The town was an early capital of Jamaica. For years Port Royal offered pirates friendly to England a safe anchorage and a marketplace for the goods they stole. In the 20 or so years before the quake, the town became somewhat more lawful—as a major trading center for sugar, raw materials, and slaves. Pirates were no longer welcome. When caught, they were hanged at Gallows Point, the peninsula at upper left.

IN CLOUDED WATER, divers examine objects on a brick floor (above) that pirates may once have paced. A corked bottle they've uncovered goes into a plastic bag weighted with a rock (inset). The weight prevents the bottle—in which a small amount of gas has formed—from floating away. Another bottle, at right, holds no gas. Bottles like the ones shown here contained wine and rum.

Robbers of the High Seas

Gold and other treasure of the New World tempted nations—and pirates. Piracy didn't begin with the discovery of the New World; sea robbers probably go as far back as ships. Still, the 16th, 17th, and 18th centuries saw much pirate activity. Some pirates—the buccaneers—took mainly Spanish targets. Others plundered ships and towns of any nation.

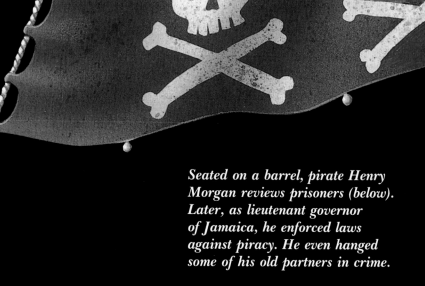

Seated on a barrel, pirate Henry Morgan reviews prisoners (below). Later, as lieutenant governor of Jamaica, he enforced laws against piracy. He even hanged some of his old partners in crime.

Wherever they roved, pirates flew flags designed to terrify. This one, displayed in the Indian Ocean by Christopher Condent, has three skulls and crossbones.

Dressed to kill: The pirate Blackbeard (left) is shown with smoking rope under his hat. It gives him a ferocious look.

Spanish ships carrying gold doubloons (left) and silver pieces of eight (above) were favorite prey of buccaneers.

Not all pirates were men. Ships flying the flag of any nation dreaded the sight of cutthroats Mary Read and Anne Bonny. Here (right) the two fight off a boarding party of the British Navy. The pirates lost. A court in Jamaica convicted and imprisoned them.

SAN ESTEBAN—1554

PORT ROYAL—

Europe's Race For Riches

Not long after the great sea ventures of Columbus, da Gama, and Magellan, the race is on to acquire the wealth of newly opened lands. Several European nations become trading rivals. The rivalry extends from the Caribbean to the Far East.

WITHIN A FEW DOZEN YEARS of Columbus's landing in the New World, Spanish and Portuguese explorers had mapped large areas of North and South America. Portugal claimed Brazil as a possession. Spain claimed much of the rest of the Americas.

In the New World, the Spanish enslaved many Indians, seized their silver mines, and took their gold. Spain began sending fleets to bring the precious metals home. It was in such a treasure fleet that *San Esteban* and three other ships (pages 64–67) sailed to their doom.

During the 1500s and 1600s, other European nations began New World settlements of their own. The English ship *Mayflower* landed in North America. The Pilgrims aboard *Mayflower* settled Plymouth, Massachusetts. It was one of the first permanent European settlements in what would become the United States.

In the Caribbean, England seized Jamaica from Spain. Port Royal became a major center in the growing trade between Europe and its colonies in the New World. Blacks brought from Africa were sold as slaves to work plantations and mines. They made up a large portion of the New World trade. The wealth crisscrossing the seas created rivalry between countries. It also attracted swarms of pirates— who sometimes did their work with the quiet approval of one government or another.

On the other side of the world, the Orient lay open for trade and often colonization. Portugal had blazed the sea road to Asia and controlled trade there for almost a century. But Portugal's hold on Asia gradually slipped. By the early 1600s a variety of European ships, including East Indiamen such as *Amsterdam* (pages 70–71), had become familiar sights in Asian ports. The Asian trade, together with the trade from the New World, brought wealth, power, and increased rivalry to the seafaring nations of Europe.

1692

AMSTERDAM—1749

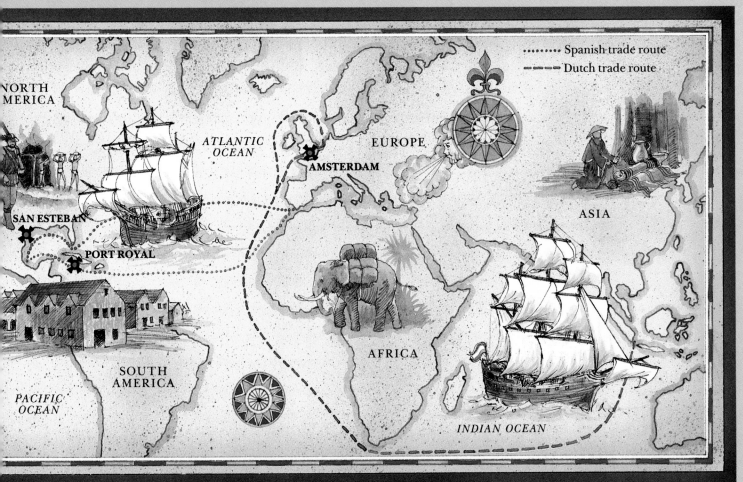

NORTH AMERICA

ATLANTIC OCEAN

EUROPE

AMSTERDAM

ASIA

SAN ESTEBAN

PORT ROYAL

AFRICA

SOUTH AMERICA

PACIFIC OCEAN

INDIAN OCEAN

········· Spanish trade route
– – – – Dutch trade route

Treasure Ships Of Padre Island

In the 16th century, as many as a hundred ships a year were leaving Spanish ports for the colonies in the New World. The ships brought the colonists such things as household items, weapons, books, wine, and letters from home. On the return trip, the ships hauled lumber, hides, dye, and—for the treasury of Spain—precious metals. On either leg of the journey, the ships faced great dangers. There were treacherous reefs, foul weather, and stony-hearted pirates. Many ships were lost.

In the spring of 1554, the small merchant ship *San Esteban* and three sister ships set out from Veracruz, Mexico, for Havana, Cuba. They were bound for Spain. They carried gold and silver weighing perhaps 85,000 pounds. They also carried passengers.

About 20 days out, a storm blew the ships off course. Three, including *San Esteban*, ran aground off Padre Island, on the coast of Texas. (The fourth ship reached Havana so badly damaged she had to be scrapped.) Only a few of the 410 or so passengers and crew from the fleet survived. Spanish authorities salvaged what they could of the cargo. They then left the ships to time and the sea.

San Esteban was rediscovered in the 1960s. Little remained of her wooden hull, but much of the cargo was intact. After mapping the site, archaeologists began recovering *San Esteban*'s scattered artifacts. Some were large: anchors, cannon, tools. Some were small: pins, coins, tableware. There was the unexpected: an Indian-made mirror of a polished mineral called pyrite. Perhaps it was a passenger's souvenir.

Searchers needed special devices to locate many of the artifacts, which lay buried in sand. Even if they had been lying in plain view, they might easily have been overlooked by divers. A rocklike crust (right) covered most of the objects. But in the laboratory, with the help of X rays and special tools, archaeologists finally saw the treasure that had lain forgotten at Padre Island for more than 400 years.

POSITION FINDER. An astrolabe (left) helped sailors calculate their latitude—distance north or south of the Equator. The instrument measured the height of the sun at noon or of stars at night. Astronomers too used astrolabes to help them study the skies (above). This artifact—oldest dated mariner's astrolabe—turned up in the wreckage of Espíritu Santo. *She was one of three Spanish merchant ships lost in a storm off Padre Island, Texas, in 1554. (A fourth ship escaped, badly battered.) Most of those aboard lost their lives. Some died in the storm or in the grounding; others were cut down by the arrows of attacking Indians.*

STOWAWAY. The wing of a New World cockroach appears, in fossil form (left), in a rocklike clump called a concretion (kon-KREE-shun). Next to it rests a present-day roach of the same kind. The fossil came from the wreck of San Esteban, *off Padre Island. (European roaches had stowed away, too.) Concretions enclosed most objects found at Padre Island, making them nearly impossible to spot with the eye alone (right). Archaeologists used special equipment to detect the treasures and took great care breaking apart the concretions.*

NUMBERED MAP of *the* San Esteban *wreck shows the locations of anchors, cannon, and other artifacts uncovered by archaeologists. Number 129 (arrow) indicates a 200-pound concretion that concealed a small wood-and-gold cross (left and inset) from which the top was missing.*

LAST-MINUTE SHOPPING. Before returning to Spain, a merchant and his family go souvenir hunting in a market in Veracruz, Mexico. The man has bought a mirror made of polished pyrite, a common mineral. In front of the family walks a young Indian girl, perhaps looking for her mother. A mule train moves through the market under armed guard. The mules carry silver bound for the treasury of Spain. It will be loaded aboard San Esteban *and three other ships anchored nearby. The ships will sail with about 410 people aboard. They represent a wide variety of ages,* *classes, and occupations. There are children and parents, soldiers and priests, traders and colonists, cabin boys and African slaves.*

HEAVY LOSS. When the ships went down, thousands of pounds of silver went with them. Bulk silver was shipped in crude disks (above) formed in sand molds. A Spanish colonial coin shows relative size. It came from a similar disk. The coin was made at the New World's first mint, in Mexico City. Stamps (below) on disks show the origin of the silver, and taxes paid on it.

All That Glitters...

Sometimes a wreck is located only after years of study and searching. Occasionally, a diver, swimmer, or beachcomber comes upon treasure by chance. The objects on these pages came from lost Spanish ships. Hundreds of other wrecks remain to be discovered.

A gold manicure set helped its owner keep groomed. It may have belonged to a ship's officer or to a wealthy passenger.

This gold cross with emeralds was a national treasure of Bermuda until 1975, when thieves stole it from a museum. It has not been recovered.

Richly decorated with a raised design, this gold tray (left) may have served to hold a nobleman's gloves.

Emeralds, amethysts, and a doubloon fill a cup-shaped gold ingot. A gold chain ten feet long surrounds the treasure.

Roman numerals on a gold bar (above) tell the gold's purity in units called karats. This gold is 22 (XXII) karats; pure gold is 24 karats.

A Tale of Storm And Mutiny

It was sewer workers who finally discovered her. Her name was *Amsterdam*. Just built, she was a sturdy merchant ship known as a Dutch East Indiaman. In January 1749, the ship got under way from the island of Texel, in the Netherlands. Her destination: the island of Java, in the Dutch East Indies (now Indonesia). The voyage was expected to last perhaps nine months. *Amsterdam* carried 28 chests of silver with which to buy porcelain ware, silk, and spices. About 54 cannon protected her.

Nothing, however, could protect her from the cruel gale that blew out of the southwest as she entered the English Channel. The storm hammered the Indiaman for more than a week. Finally the captain, Willem Klump, headed for a sheltered bay. But the rudder was torn off and *Amsterdam* drifted out of control. Klump dropped anchor off Hastings, in England. The gale continued without letup. The crew mutinied and forced Captain Klump to beach the ship.

Those aboard waded ashore. The silver was salvaged—except for one chest that was broken into by a band of English smugglers (right). Then the ship began to sink into the sand and clay.

Exactly 220 years later, in 1969, workers were building a sewer outlet off Hastings beach. On a rest break, they decided to dig at the wreck site with their modern excavating equipment. The timbers had always been visible at extremely low tide, but they had been largely ignored. When the workers dug up cannon and old wine bottles, they knew they'd hit on something big. They turned the site over to archaeologists. Slowly, carefully, the scientists investigated the wreck, uncovering rare artifacts. The chief archaeologist, Peter Marsden, calls *Amsterdam* one of Europe's most important finds—"an almost untouched storehouse of life in the 18th century."

AT LOW TIDE, the hull of the wrecked merchant ship Amsterdam *lies revealed on the coast of England (above). In 1749, on her first voyage, this Dutch East Indiaman ran into a savage gale and lost her rudder. The crew mutinied, demanding that the ship be grounded. She eventually sank into the seabed and was largely forgotten. Here, archaeologists survey the wreck.*

NOW CRUSTED OVER, a cannon (below) once helped defend Zeewijk, *a Dutch East Indiaman that hit a shallow reef off Australia in 1727.* *Because the big trading ships often had to fight off pirates (and warships from rival countries), they carried many heavy guns.*

IN THE DARK OF NIGHT, thieves help themselves to silver ingots aboard the beached Amsterdam. Smuggler Anthony Watson, at center, leads the group. While the ship's captain consulted with officials in the nearby town of Hastings, the thieves sneaked onto the ship and broke into the captain's cabin. There the precious cargo lay in neat stacks inside locked trunks. Watson and his men got away with the entire contents of one trunk: 50 ingots weighing about 5 pounds each. They left untouched 27 other trunks. The silver was to be traded in the Dutch East Indies for Oriental goods, such as silk and spices. Two porcelain plates (right) were made in China especially for the Dutch East India Company (Vereenigde Oostindische Compagnie—VOC). A trading giant, it owned Amsterdam and hundreds of other ships.

STRANGERS IN TOWN. A section of a 16th-century Japanese screen shows Portuguese traders and missionaries arriving at a port city in Japan. From shops that line the city's main street, residents observe the scene, which the artist has greatly exaggerated: A slave holds an umbrella to shade a ship's captain. Merchants in baggy pantaloons walk pampered dogs. The

foreigners, in the eyes of the Japanese, are all remarkably tall, and all have comically long noses.

Among the first Europeans to reach Asia by sea, the Portuguese were perhaps the first to visit Japan. They arrived from the south. Because of their odd ways, the Japanese called them nanban-jin—"barbarians from the south."

In time, other countries entered the Asian trade. The goods taken home by merchants altered everyday life in a number of ways. Britain became a nation of tea drinkers. Cheap Indian cotton introduced large numbers of Europeans to soft undergarments— blessed relief from scratchy wool clothes. New groups of people could now afford to enjoy spicy food and to buy the works of Oriental craftsmen.

GEMS CALLED AGATES,
FROM BRITISH
EAST INDIAMAN

73

5

Lost Ships of Frontier America

Struck by red-hot cannonballs, H.M.S. *Charon* burns in the York River, at Yorktown, Virginia. A mast with its rigging blazes and crackles like a giant sparkler. On board, crewmen throw water on the flames in a desperate attempt to put out the fire. Their efforts will fail. *Charon*, a British warship, will sink, in the last major campaign of the American Revolution.

75

Fiery Sinking At Yorktown

The war was not going well for General George Washington. Six years after the American Revolution had begun, Britain still had the upper hand. The new nation had run out of money. Morale among the American troops was low. Washington needed a major victory.

He found that victory—and won the war—at Yorktown. Yorktown, once a major tobacco port, lies 6½ miles up the York River from Chesapeake Bay. In August 1781, the British commander, Lord Cornwallis, landed his army at Yorktown. He intended to establish a safe port for British warships. By September, however, his supplies were running low. He was also short of men. Anxiously, he awaited reinforcements from the British fleet.

But a French fleet arrived first. France was an American ally, and the French fleet drove off the approaching British warships (map, below). Meanwhile, George Washington was leading an army toward Yorktown. By September, Washington's French and American soldiers had cut Yorktown off by land. Cornwallis was trapped.

To prevent an attack by water, he ordered a number of British

MELTED DOORKNOB
FROM *CHARON*

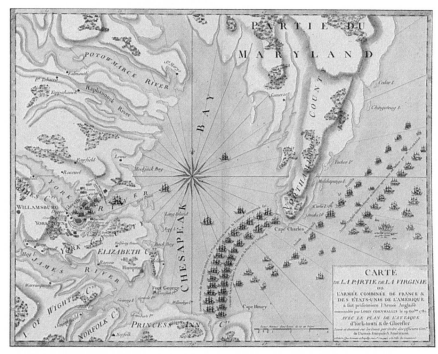

NO AMERICAN SHIP FOUGHT in the Revolution's most important naval battle. In the Battle of the Capes, in 1781, a French fleet drove off a British fleet at the mouth of Chesapeake Bay. The French then blockaded the bay. The action, shown on a French map (above), sank British hopes for more men and for supplies. It enabled General George Washington, leading a combined force of French and American troops, to seize Yorktown and win the last major battle of the war.

transport ships sunk along the shoreline to form a barricade. Washington began bombarding Yorktown. The evening of October 10, some hot shot—cannonballs heated in a furnace—hit Cornwallis's largest warship, H.M.S. *Charon*, setting her ablaze. She crashed into at least two other ships, igniting them. "The ships were enwrapped in a torrent of fire," an American observer wrote. "All around was thunder and lightning from our numerous cannons and mortars." *Charon* burned all night and sank near the opposite shore.

Cut off by land and sea, and under constant cannon fire, Lord Cornwallis decided his position was hopeless. On October 19, 1781, he surrendered his army to General Washington.

In 1973 the area where the Yorktown wrecks lie was declared a place of national historical importance. The remains of a number of ships rest there. Archaeologists have examined nine so far, including *Charon*'s charred hull. Artifacts from the ships and from the land battle are displayed at the Yorktown battlefield. The artifacts give people a rare trip back to the time and place where America fought her last big battle for independence.

SHIPSHAPE. Kisha Temple, 12, of Hampton, Virginia, examines a hammock on a reproduction of part of Charon's *gun deck. Charles Rimmer, 9, of Yorktown, peers out a door. To make this reconstruction, builders at the Yorktown Visitor Center studied old records. At least nine, and possibly many more, British ships lie at the bottom of Yorktown harbor. The British commander, Lord Cornwallis, sank some of them to create a barricade against a possible enemy landing.*

*IN NEAR
DARKNESS,
crewmen aboard a British transport
take their mess rations—that is, eat a
meal (above). The cook ladles out
watery beef soup to the ship's boy. The
meal also includes peas and hard biscuits called hardtack. As the men eat,
rats scurry about, searching for scraps.
A duck—perhaps a pet, perhaps a
future meal—eyes the leftovers. Here*

*on the lower deck, only a short man
can stand without stooping. Barrels
stacked along the bulkhead, or wall,
make conditions even more cramped.
The barrels contain water, rum, and
various foods. The darkness helps
hide the worms and bugs that crawl
in the crewmen's food. This painting
is based on artifacts found during the
excavation of a British transport at
Yorktown. The ship was probably a*

*merchant vessel pressed into
wartime British naval service.*

*SPRUNG APART after two centuries
in the York River, barrels from a
British transport stand ready for
removal (right). Tags on each piece
will help experts reassemble the
barrels. Archaeologists nicknamed
the barrels Cornwallis's hot tubs
because of their large size.*

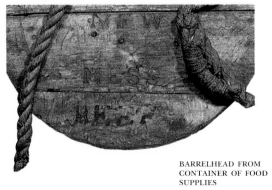

BARRELHEAD FROM
CONTAINER OF FOOD
SUPPLIES

CHARON—1781

Growing Pains Of a New Nation

America's victory in the Revolutionary War is followed by a series of revolutions elsewhere. The United States expands, soon reaching the Pacific Ocean. That expansion contributes to a bitter dispute between North and South.

IN THE LATE 1700s, Britain needed money to defend her growing empire. To obtain it, King George III and Parliament increased taxes on the American colonies. Many colonists were outraged. When their complaints were ignored, the colonists declared independence.

The American Revolution lasted eight years. The outcome was decided in 1781 at Yorktown, Virginia. A French naval blockade and George Washington's army defeated a major British force there. During the battle, H.M.S. *Charon* and a number of other British ships went down.

The 1783 treaty that actually ended the war gave the United States her independence. It also turned over a huge expanse of land between the Atlantic Ocean and the Mississippi River.

The American Revolution helped touch off revolutions elsewhere. In 1789, the French middle class overthrew the rule of the nobility and set up a new, if shaky, government. Between 1810 and 1824, most of the Spanish and Portuguese colonies in Latin America won their independence.

Meanwhile, the United States fought another war with Britain. The War of 1812 erupted over British seizures of American ships and sailors. During this war, which ended in a truce, Francis Scott Key wrote "The Star-Spangled Banner," and the American ships *Scourge* and *Hamilton* (pages 82–83) were lost.

America expanded westward, fueling a bitter debate in the East between northern and southern states. A major issue: Should slavery expand west with the nation? That dispute, with others, led in 1861 to the Civil War.

During the war, gold was discovered in the Montana Territory. The riverboat *Bertrand* (pages 84–87) was one of hundreds carrying people and goods to the goldfields. She sank in 1865, eight days before the end of the Civil War—and of slavery in the United States.

HAMILTON—1813

BERTRAND—1865

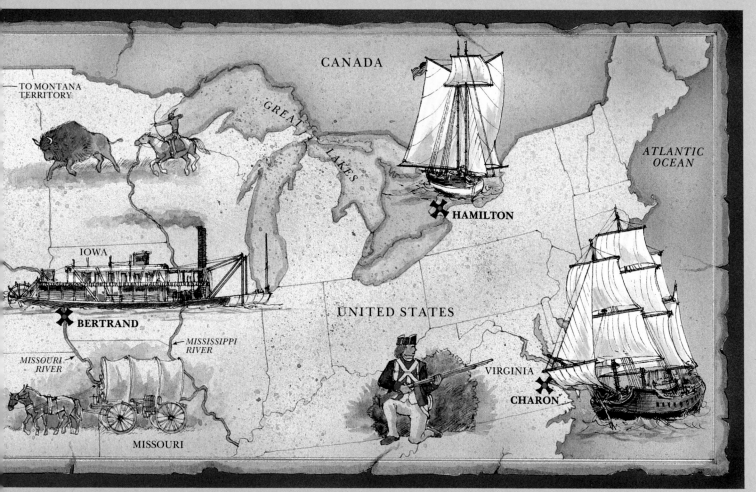

TO MONTANA TERRITORY

CANADA

ATLANTIC OCEAN

GREAT LAKES

IOWA

✠ **HAMILTON**

✠ **BERTRAND**

UNITED STATES

MISSISSIPPI RIVER

MISSOURI RIVER

VIRGINIA

✠ **CHARON**

MISSOURI

81

Ghostly Find In Lake Ontario

"The flashes of lightning . . . nearly blinded me. Our decks seemed on fire, and yet I could see nothing. . . . the schooner was filled with the shrieks and cries of the men. . . . I made a spring and fell into the water. . . . the schooner sunk as I left her."

That's how seaman Ned Myers, in a popular retelling, described his escape from the schooner *Scourge* on Lake Ontario. *Scourge* was a merchantman armed for U. S. Navy service during the War of 1812. Early on the calm morning of August 8, 1813, *Scourge* and 12 other ships lay waiting for dawn. At sunup, they were to fight a squadron of British ships. About 2 o'clock, a sudden storm caught *Scourge* by surprise. Strong winds heeled her over hard, and she capsized. A sister ship, *Hamilton*, also sank. Myers escaped, saving 7 others, as he told it. In all, 19 survived, and 53 died.

The ships lay largely forgotten for the next 158 years. Then, in 1971, the Royal Ontario Museum at Toronto, in Canada, set out to search for the wrecks. A team located the site in 1973. The ships were positively identified in 1975. Later, a robot craft was sent down to explore and photograph the site.

The robot recorded a ghostly scene. The ships lie intact, preserved by the lake's cold waters and by the darkness at 300 feet. Cannon point through open gunports. Cannonballs sit neatly in their racks. Swords called cutlasses stand ready to be grabbed. Around the wrecks lie the skeletons of drowned crewmen.

Specialists are studying the ships to decide if they can be raised. The city of Hamilton, Ontario, which owns the wrecks, hopes it's possible. Hamilton has set aside a lakeside area where a museum would be built to house these ships and to help preserve their rich history.

DEEP IN ICY WATERS, an anchor from the schooner Hamilton *hangs from a timber (above). A sudden early morning storm sent* Hamilton *to the bottom of Lake Ontario in Canada during the War of 1812.*

SONAR IMAGES such as this one (left) helped archaeologists find Hamilton *and her sister ship* Scourge, *which also sank in the storm. The black image shows the ship; white shows the "sound shadow" the ship casts on the lake bottom. Sonar images record reflected sound waves, much as photographs record reflected light.*

ABOUT 300 FEET UNDER, a remote-control camera took this picture (right) of Hamilton's *figurehead, the ancient goddess Diana. The ship, on temporary duty with the U. S. Navy, sank while awaiting battle with a British squadron.*

A Riverboat's April Fools' Day

LABEL FROM CAN OF POWDERED LEMONADE

BOTTLE OF BRANDIED PEACHES

Perhaps 30 passengers rode the steamboat *Bertrand* as she made her way up the Missouri River. The boat had left St. Louis, Missouri, March 18, 1865. She was headed for the goldfields of the Montana Territory. John Walton, 11, was aboard. So were teenage sisters Fannie and Annie Campbell, on their way home from boarding school.

Two weeks into the journey, in the afternoon of April 1, *Bertrand* hit a submerged tree that ripped open the hull. John was on the top deck—although it was off-limits to passengers. He leaped into the water and swam to shore. The pilot, Horace Bixby, guided the sinking stern-wheeler close to the riverbank. All on board escaped unharmed.

Bertrand eventually settled into the river bottom. The Missouri changed course, leaving the boat's exact whereabouts unknown. In 1967, two salvors, working under an agreement with the U. S. Government, set out to find the boat. Using old records and modern detection equipment, they did find *Bertrand*. She was buried 30 feet deep in a field. The site is in a national wildlife refuge that covers parts of Iowa and Nebraska.

Fannie Campbell's chalkboard was one of the tens of thousands of items recovered. The name Fannie was etched into the frame. Excavators found the types of frontier cargo you'd expect. There were pickaxes and woolen underwear, sleigh bells and plows, building supplies and butter churns, catsup and soda crackers.

There were also surprises. *Bertrand* carried 780 gallons of Dr. J. Hostetter's Celebrated Stomach Bitters, a popular, highly alcoholic cure-all. The cargo included fine men's suits, silk ribbon, champagne, French olive oil, brandied peaches, and canned oysters. Clearly, life on the frontier was not harsh for everyone.

STUCK IN THE MUD, the steamboat Bertrand *sits in a former channel of the Missouri River (left).* Bertrand *sank here on April Fools' Day 1865, after hitting a snag. In this picture, archaeologists inspect the hull during excavation in 1969. The cargo included more than 6,000 bottles. Most still held their original contents.*

HOW WOULD IT TASTE? Andrew Schultz, 9 (right), wonders. These bottles and other artifacts from Bertrand *are displayed at the DeSoto National Wildlife Refuge, in Missouri Valley, Iowa. That's Andrew's hometown.*

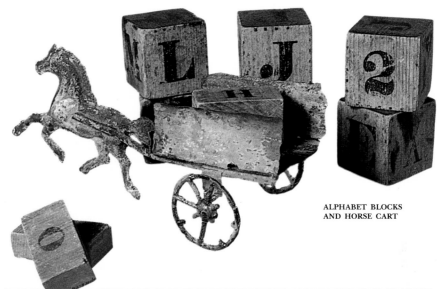

ALPHABET BLOCKS
AND HORSE CART

LIFE ON THE MISSOURI. Bertrand *carried a cargo of foodstuffs, tools, and some luxury items for the gold-mining communities of the Montana Territory. She also carried passengers, perhaps 30 in all. The toys on this page probably belonged to passengers Charles Atchison, 5, and his sister, Emma, 4.*

As Bertrand *churned her way upriver (left), the children spent the time playing, watching for Indians, taking reading lessons, and enjoying the wild scenery. Sundays they attended church services on board.*

Bertrand *sank two weeks into the trip, 25 miles north of Omaha, Nebraska. All aboard got off safely. One passenger, 11-year-old John Walton, grew up to become a gold miner in the Montana Territory. Another passenger, teenager Fannie Campbell, became a homesteader there.*

YOUNGSTER'S SHORT
PANTS; CHILDREN'S
BOOTS

BUILDING BLOCKS

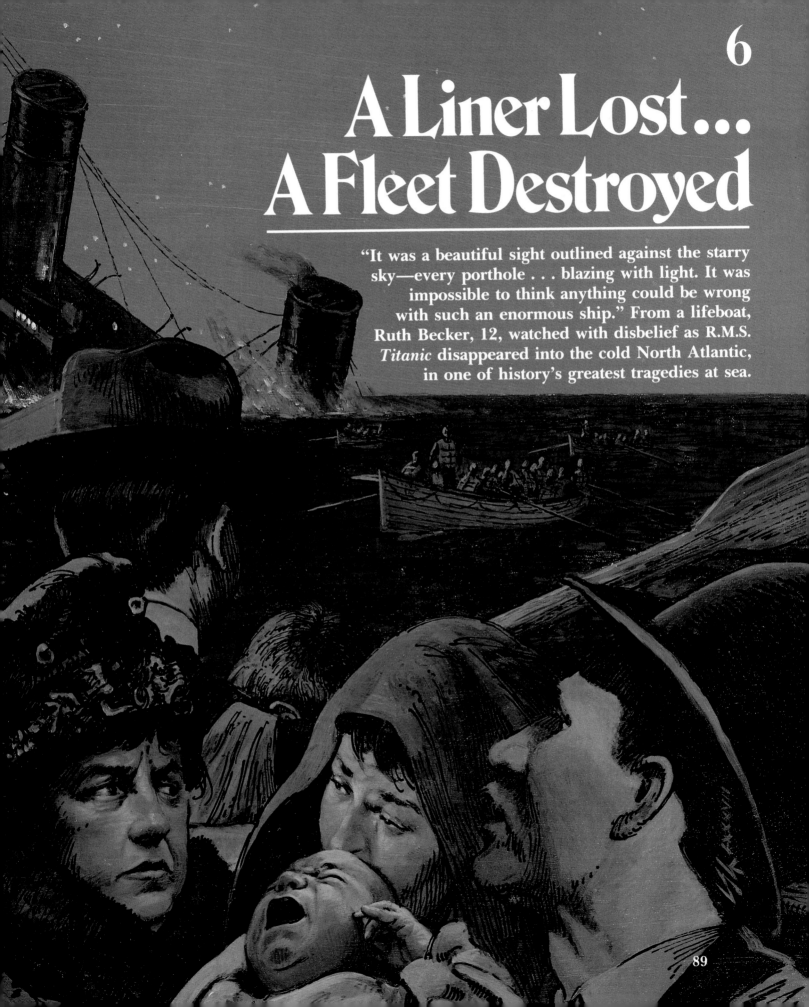

6

A Liner Lost...
A Fleet Destroyed

"It was a beautiful sight outlined against the starry sky—every porthole . . . blazing with light. It was impossible to think anything could be wrong with such an enormous ship." From a lifeboat, Ruth Becker, 12, watched with disbelief as R.M.S. *Titanic* disappeared into the cold North Atlantic, in one of history's greatest tragedies at sea.

Titanic's Doomed Voyage

She was the largest ocean liner yet built when she steamed out of Southampton, England, on April 10, 1912, to begin her maiden voyage across the Atlantic to New York. A sixth of a mile long and 150 feet from waterline to the top of her smoke-belching funnels, the luxury liner R.M.S. *Titanic* was truly a floating city. Her 2,228 passengers and crew traveled in comfort and security—until the fourth day at sea.

It was almost midnight. The sea lay flat. The air was cold and clear, except for a slight haze dead ahead. *Titanic* cut through the water at 21½ knots, almost 25 miles an hour. Suddenly, she gave a shudder. It felt, a passenger later said, "as though somebody had drawn a giant finger along the side of the ship." *Titanic* had sideswiped an iceberg hidden in its own haze. The berg opened up *Titanic*'s hull below the waterline. Water rushed in.

The ship was going down, and Captain Edward J. Smith knew it. Smith instructed his officers to start loading the lifeboats. But *Titanic*, considered unsinkable, carried far too few of them. In the confusion, some were lowered with fewer people than they could hold. As the last lifeboat pulled away, more than 1,500 people remained behind. All of them perished as *Titanic* disappeared beneath the sea.

A full 73 years were to pass before the ship would be seen again. In 1985, a team of French and American scientists located the wreck. In 1986, the Americans carefully explored the ghostly scene. The ship lies more than 2 miles down, 350 miles off Newfoundland. She will probably never be raised. Many would oppose such an effort. Wrote survivor Ruth Becker Blanchard: "I hope [*Titanic*] will be left in peace down there—as a quiet resting place for those who were not so fortunate, in those terrible hours, as I was."

A RUST FORMATION grows over one of Titanic's *portholes. The ship lay hidden for 73 years before being found in 1985. She sank in 1912 on her first voyage, while sailing from England toward New York.*

DEEP BELOW THE SURFACE, special lights and cameras help a robot named Jason Jr. *(J. J., for short) peer into a first-class cabin. Some first-class passengers, seeking maximum luxury, paid as much as $50,000 (in today's dollars) for the expected six-day passage. The heavy loss of life that followed* Titanic's *collision with an iceberg could have been prevented. There were several ships in the general area, but they were unaware of* Titanic's *distress calls. All the passengers could have escaped in lifeboats, but the "unsinkable"* Titanic *carried too few. Of the 2,228 people aboard, only 705 survived.*

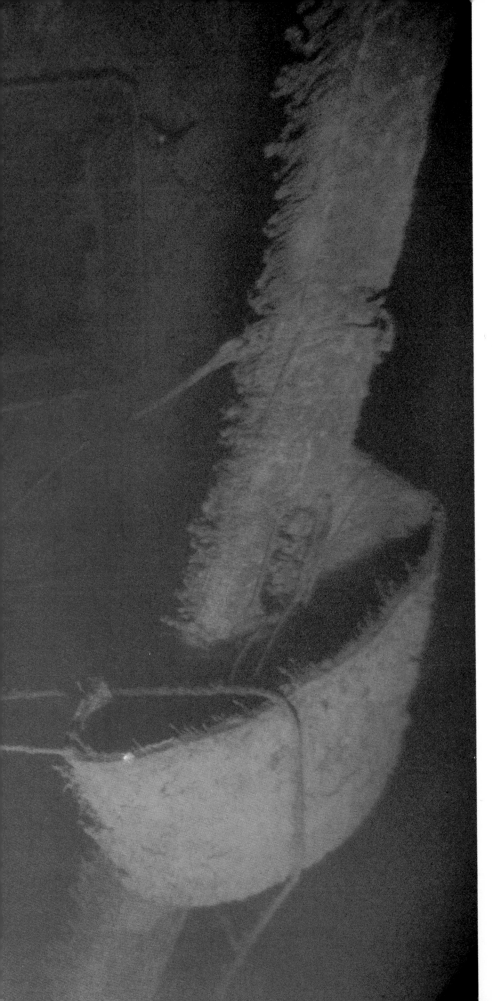

FROM THE CROW'S NEST (left) on Titanic's *forward mast, lookouts kept watch on the sea—and first spotted the fatal berg. In 1986, the robot* J. J. *explored* Titanic *with a partner,* Alvin, *a three-person submersible. In the cutaway diagram above,* J. J., *attached to* Alvin *by a 200-foot tether, explores the ship's interior. Only 28 inches long,* J. J. *can go into areas that are too small or too dangerous for* Alvin *and crew. A team led by Dr. Robert D. Ballard undertook the* Titanic *project. Dr. Ballard is a marine geologist at the Woods Hole Oceanographic Institution, in Woods Hole, Massachusetts.*

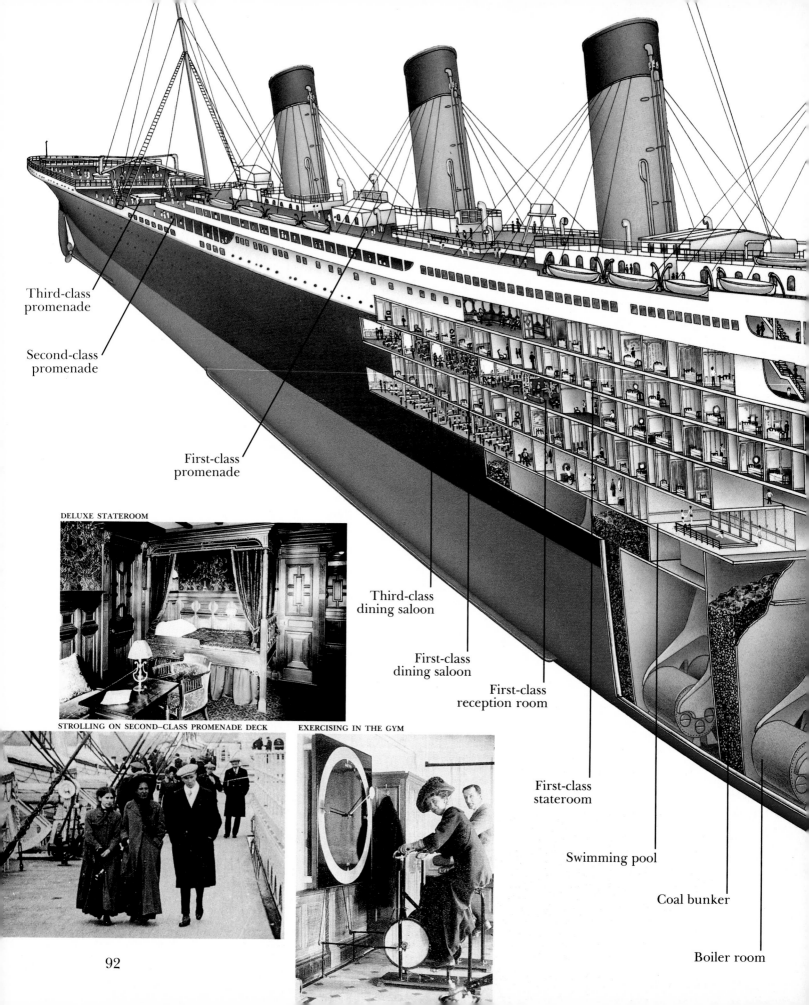

Third-class
promenade

Second-class
promenade

First-class
promenade

DELUXE STATEROOM

Third-class
dining saloon

First-class
dining saloon

First-class
reception room

First-class
stateroom

Swimming pool

Coal bunker

Boiler room

STROLLING ON SECOND-CLASS PROMENADE DECK

EXERCISING IN THE GYM

92

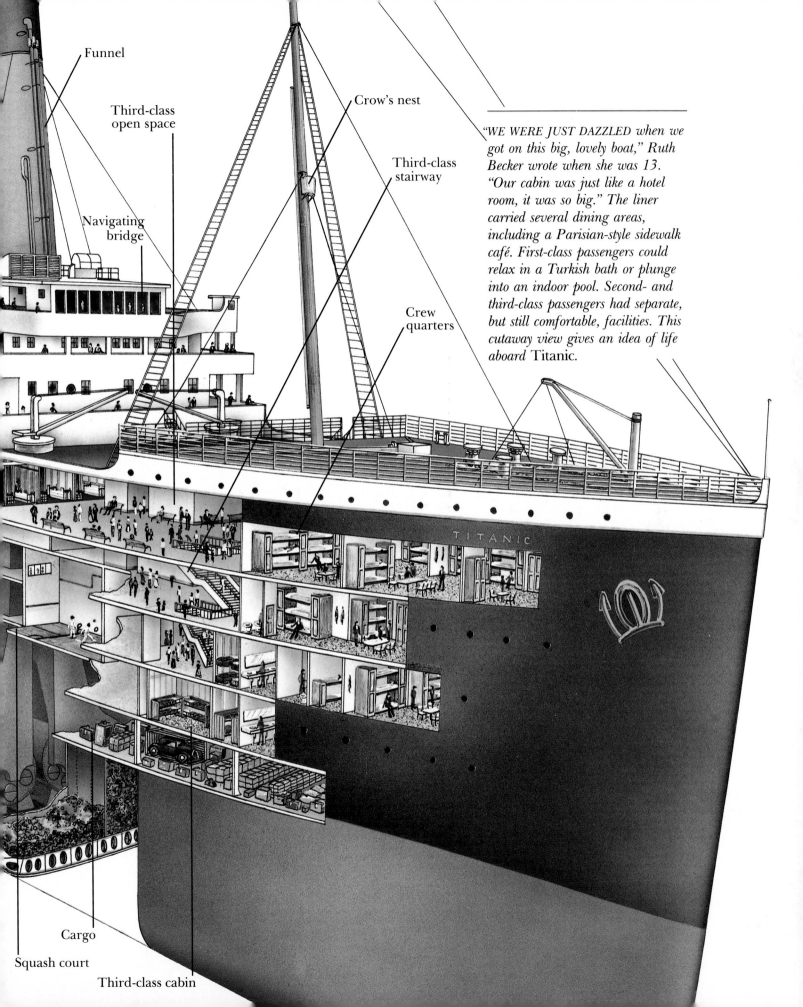

Funnel

Third-class
open space

Crow's nest

Third-class
stairway

Navigating
bridge

Crew
quarters

"WE WERE JUST DAZZLED when we got on this big, lovely boat," Ruth Becker wrote when she was 13. "Our cabin was just like a hotel room, it was so big." The liner carried several dining areas, including a Parisian-style sidewalk café. First-class passengers could relax in a Turkish bath or plunge into an indoor pool. Second- and third-class passengers had separate, but still comfortable, facilities. This cutaway view gives an idea of life aboard Titanic.

TITANIC

Cargo

Squash court

Third-class cabin

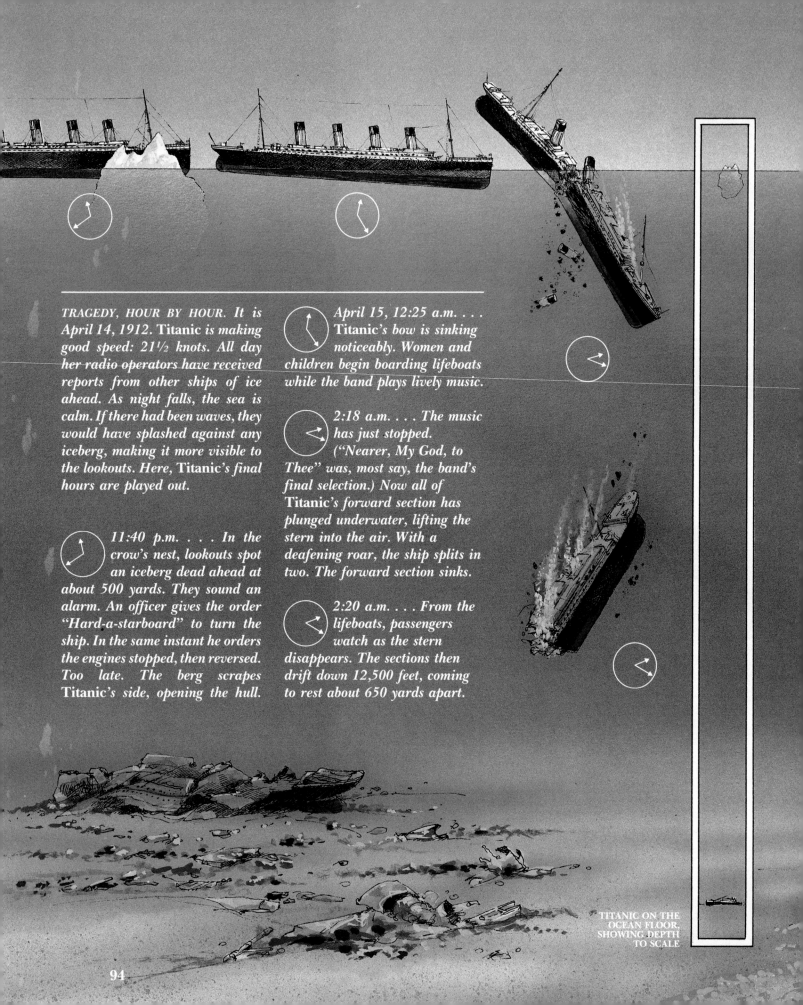

TRAGEDY, HOUR BY HOUR. *It is April 14, 1912.* Titanic *is making good speed: 21½ knots. All day her radio operators have received reports from other ships of ice ahead. As night falls, the sea is calm. If there had been waves, they would have splashed against any iceberg, making it more visible to the lookouts. Here,* Titanic's *final hours are played out.*

11:40 p.m. . . . In the crow's nest, lookouts spot an iceberg dead ahead at about 500 yards. They sound an alarm. An officer gives the order "Hard-a-starboard" to turn the ship. In the same instant he orders the engines stopped, then reversed. Too late. The berg scrapes Titanic's *side, opening the hull.*

April 15, 12:25 a.m. . . . Titanic's *bow is sinking noticeably. Women and children begin boarding lifeboats while the band plays lively music.*

2:18 a.m. . . . The music has just stopped. ("Nearer, My God, to Thee" was, most say, the band's final selection.) Now all of Titanic's *forward section has plunged underwater, lifting the stern into the air. With a deafening roar, the ship splits in two. The forward section sinks.*

2:20 a.m. . . . From the lifeboats, passengers watch as the stern disappears. The sections then drift down 12,500 feet, coming to rest about 650 yards apart.

TITANIC ON THE OCEAN FLOOR, SHOWING DEPTH TO SCALE

UNDERWATER MUSEUM. Calling up visions of "those terrible hours," objects from **Titanic** lie scattered across the ocean floor. The head of a ceramic doll (upper left) still wears a smile. The wooden slats on a deck bench (center) are gone, possibly eaten away by undersea creatures. A teak staircase (upper right) remains largely intact. Chemicals applied to keep the wood from weathering may have preserved it. A porcelain toilet bowl (lower right) gleams white beside a bottle. After nearly three quarters of a century underwater, a copper pot (lower left) still glimmers. Thousands of artifacts like these, found near **Titanic's** stern section, gave scientists an eerie look at life, and death, aboard the ill-fated luxury liner.

TITANIC—1912

20th Century: Full Speed Ahead

The world has changed tremendously in the past hundred years—more than in any other century in history. Now the world faces a vital challenge.

WHEN THIS CENTURY BEGAN, the Wright brothers had not yet made their historic flight. Ships were the only means of carrying people and goods across the oceans. The automobile was not in general use. Trains carried people long distances over land. As for communications, the telephone was a new and little-used gadget. No one in the world had heard a radio program— much less watched television.

In world politics, kings and queens in Europe still ruled over vast colonial empires in Asia and Africa. The colonies supplied Europe with cheap raw materials and with a ready market for finished goods.

Since 1900, developments in science and technology have changed, almost beyond imagination, the way we live. In the late 1800s, for example, ships— sailing ships—were an essential form of transport. Sail gave way to steam and to such floating cities as the liner R.M.S. *Titanic*. Now jet planes whisk travelers from continent to continent in a matter of hours, rather than the days or weeks that steamships took.

Wars and world politics too have changed the world and our view of it. This century has seen the destruction brought by two world wars and many smaller ones. World War I began only 2 years after the *Titanic* disaster. Just 21 years after that war ended, World War II began. The lives lost at Truk (pages 98–100) accounted for only a tiny percentage of the millions killed.

The drive for political freedom has continued strong in this century. Dozens of former colonies have become nations. Sometimes they were given independence; often they had to fight for it.

Greatly improved transportation and communications have brought the world's people closer together. At the same time, nuclear weapons have made global war unthinkable. People today face the most important challenge in history: living together in peace on this small planet.

TRUK LAGOON FLEET—1944

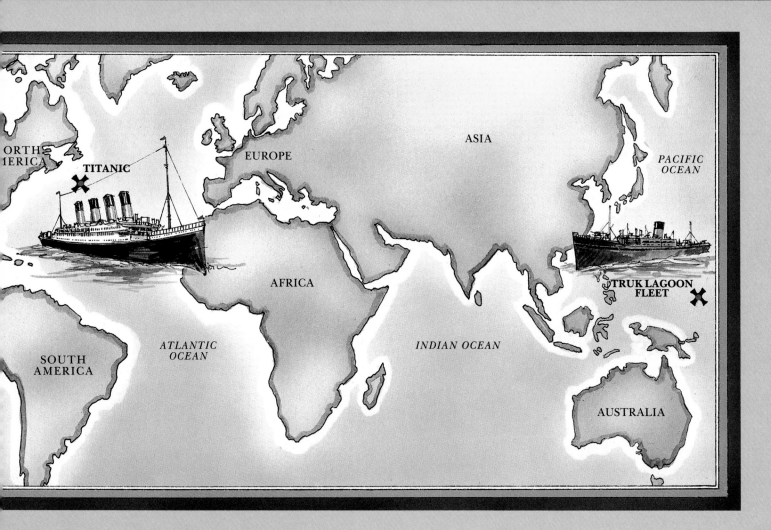

Battle at Truk Lagoon

I f you were looking for the ideal spot to build a secret naval base, what kind of place would you choose? About 50 years ago, the Japanese picked Truk, a cluster of islands in the western Pacific Ocean. A 125-mile coral reef ringed the islands, protecting them from storms. Four channels gave ships passage into a large natural harbor, Truk Lagoon. Truk, though in a little-visited area, was within easy reach of much of the Pacific.

The Japanese built Truk into an important base. World War II had not yet begun, and the United States had little interest in the islands. (One admiral was to remark that all he knew about Truk was what he read in NATIONAL GEOGRAPHIC.) Then, in a surprise attack on December 7, 1941, Japan bombed the U. S. naval base at Pearl Harbor, in Hawaii. The United States immediately declared war. Naval officers dusted off their maps of Truk.

Twenty-six months after the attack on Pearl Harbor, early on February 17, 1944, the sky northeast of Truk filled with U. S. Navy Hellcat fighters. U. S. planes battered enemy aircraft and pounded airfields. They wiped out ground installations and sank all ships in the lagoon. After repeated attacks, Truk's days as a naval base were over.

Today, Truk Lagoon attracts scuba divers from all over the world. The lagoon holds some 60 ships, now covered by marine growth. They include destroyers, troop carriers, tankers, freighters. Aircraft—dozens of them, from both sides—litter the bottom. There are big guns, ammunition, and human skeletons. Marine life of every description thrives in and around the wreckage—the world's largest collection of artificial reefs. Divers may explore the sunken hulks of Truk, but only the sea has claim to them. The lagoon has been designated a historical monument, preserved for future generations.

CLEAR WATERS of Truk Lagoon reveal gas masks lying in the wreckage of a Japanese warship. The ship was sunk during a U. S. attack on the Pacific base during World War II.

SMOKE BILLOWS SKYWARD as bombs fall during Operation Hailstone. That was the U. S. Navy's code name for its 1944 attack on Truk. This Japanese naval base served as a fleet headquarters and as an anchorage. The attack sank all the ships at Truk, about 60 of them. The ships became artificial reefs, and today the lagoon attracts sight-seeing sport divers. "The sea is healed," one islander observed.

HOLDING A FLASHLIGHT, a diver inspects a Japanese tank. It rests on the deck of San Francisco Maru, an armed transport.

Porkfish glide around a shipwreck off the coast of Florida. This ship was torpedoed by a German submarine in World War II.

Adding their bright colors to the underwater world, hydrocorals—colonies of tiny marine animals —grow on a shipwreck in Truk Lagoon.

Living Treasures Under the Sea

Shipwrecks attract many kinds of sea creatures—and scuba divers as well. This small freighter, *Oro Verde*, was intentionally sunk in 1980 in the Caribbean Sea to create an artificial reef. Archaeologists find recent wrecks less interesting than older ones. The scientists already know about modern customs and shipbuilding methods. Still, by supporting a world of life and color, new wrecks create their own hidden treasures of the sea.

Index

Consultants

George F. Bass, Archaeological Director, Institute of Nautical Archaeology; and Abell Professor of Nautical Archaeology, Texas A&M University

J. Richard Steffy, Ship Reconstructor, Institute of Nautical Archaeology; and Professor of Nautical Archaeology, Texas A&M University

Joan W. Myers—*Reading Consultant*

Nicholas J. Long, Ph.D.—*Consulting Psychologist*

The Special Publications and School Services Division is grateful to the individuals named or quoted within the text and to those cited here for their generous assistance:

Martin P. Amt, Freer Gallery of Art; J. Barto Arnold III, Texas Antiquities Committee; Elizabeth Bailey, Missouri Historical Society; Robert H. Brill, Corning Museum of Glass; John Broadwater, Virginia Department of Conservation and Historical Resources; Emily Smith Cain, Hamilton–Scourge Foundation; Tomoko Campen, Arlington, Virginia; Timothy J. Carter, National Geographic Society; Valerie Chase, National Aquarium in Baltimore; Geoffrey Chester, Albert Einstein Planetarium; Piero Cortese, Archeo Tecnica Mare, Rome, Italy; Ole Crumlin–Pedersen, Institute of Maritime Archaeology, The Danish National Museum; Ted Dachtera, National Geographic Society.

Roberta D'Amico, Colonial National Historical Park; John L. Davis, The University of Texas at San Antonio; John P. Eaton, Titanic Historical Society; Robert Grenier, Canadian Park Service; Charles A. Haas III, Titanic Historical Society; D. L. Hamilton, Texas A&M University; James N. Haskett, Colonial National Historical Park; Robert L. Hohlfelder, University of Colorado, Boulder; Donald H. Keith, Institute of Nautical Archaeology; Emory K. Kristof, National Geographic Society; Shelley M. Lauzon, Woods Hole Oceanographic Institution; James B. Levy, Florida Division of Historical Resources.

Peter Marsden, Museum of London; James J. Miller, Florida Division of Historical Resources; James P. Murray, St. Timothy's School, Stevenson, Maryland; William M. Murray, University of South Florida, Tampa; Alessandra Negriolli–Kopping, Greenville, North Carolina; Leslie A. Perry Peterson, DeSoto National Wildlife Refuge; Jerry Petsche, National Park Service; Cemal Pulak, Institute of Nautical Archaeology; John C. Reilly, Naval Historical Center; Marcie Renner, Virginia Department of Conservation and Historical Resources; Joseph Ringel, The National Maritime Museum, Haifa, Israel; Margaret Rule, The Mary Rose Trust; John O. Sands, The Colonial Williamsburg Foundation; Tom F. Shelton, The University of Texas Institute of Texan Cultures.

Anita Solak, The Hispanic Society of America; Robert Stenuit, Brussels, Belgium; Willis Stephens, Canadian Park Service; Susan van Hoek, Harbor Branch Oceanographic Institution, Inc.; R. Lindley Vann, University of Maryland; Katarina Villner, Sjöhistoriska Museet, Wasavarvet, Stockholm, Sweden; Robert L. Webb, Kendall Whaling Museum; John Sampson White, National Museum of Natural History.

Additional Reading

Readers may want to check the *National Geographic Index* or a public library for related articles and to refer to the following books ("A" indicates a book for readers at the adult level.)

Abranson, Erik, *Sailors of the Great Sailing Ships*, Flammarion, 1977. Abranson, Erik, *Ships and Seafarers*, Silver Burdett, 1980. Ballard, Robert D., *The Discovery of the Titanic*, Warner Books, 1987 (A). Bass, George F., (ed.), *A History of Seafaring Based on Underwater Archaeology*, Walker, 1972 (A). Bass, George F., (ed.), *Ships and Shipwrecks of the Americas: A History Based on Underwater Archaeology*, Thames & Hudson, 1988 (A).

Cain, Emily, *Ghost Ships: Hamilton and Scourge—Historical Treasures From the War of 1812*, Beaufort Books, 1983 (A). Civardi, Anne, and James Graham–Campbell, *The Time Traveller Book of Viking Raiders*, Hayes Books, 1977. Davis, John L., *Treasure, People, Ships, and Dreams*, Texas Antiquities Committee Publication No. 4 and The Institute of Texan Cultures, 1977.

Eaton, John P., and Charles A. Haas, *Titanic: Triumph and Tragedy*, W. W. Norton, 1986 (A). Ford, Barbara, and David C. Switzer, *Underwater Dig: The Excavation of a Revolutionary War Privateer*, William Morrow, 1982. Franzen, Greta, *The Great Ship Vasa*, Hastings House, 1971. Grant, Neil, *The Discoverers*, Arco, 1979. Landström, Björn, *Sailing Ships*, Doubleday, 1969. Landström, Björn, *The Ship*, Doubleday, 1961. Landström, Björn, *The Warship Vasa*, Interpublishing, 1988.

Marsden, Peter, *The Wreck of the Amsterdam*, Hutchinson, 1974 (A). Muckelroy, Keith, (ed.), *Archaeology Under Water: An Atlas of the World's Submerged Sites*, McGraw–Hill, 1980 (A). Naish, G. P. B., and Heather Amery, *The Age of Sailing Ships*, Usborne Publishing, 1976. Petsche, Jerome E., *The Steamboat Bertrand*, National Park Service, 1974 (A). Preston, Anthony; David Lyon; and John H. Batchelor, *Navies of the American Revolution*, Prentice–Hall, 1975. Rogers, Cedric, *Sailing Ships*, Golden Press, 1974. Rule, Margaret, *The Mary Rose*, Conway Maritime Press, 1982 (A).

Sands, John O., *Yorktown's Captive Fleet*, University Press of Virginia, 1983 (A). Throckmorton, Peter, (ed.), *The Sea Remembers*, Weidenfeld and Nicolson, 1987 (A). Time–Life Books, The Seafarers (series), beginning 1978 (A).

BOOKS BY THE NATIONAL GEOGRAPHIC SOCIETY: *The Adventure of Archaeology*, 1985. *Exploring the Deep Frontier: The Adventure of Man in the Sea*, 1980. *Men, Ships, and the Sea*, 1973. *Romance of the Sea*, 1981.

Library of Congress CIP Data

Hidden treasures of the sea.

(Books for world explorers)

SUMMARY: An introduction to nautical archaeology, focusing on social, cultural, and political history as exposed by shipwrecks from ancient times to the present.

1. Treasure trove—Juvenile literature. [1. Shipwrecks.
2. Underwater archaeology. 3. Archaeology] I. National Geographic Society (U.S.) II. Series.

G525.H474 1988   910.4'53   88-5142

ISBN 0–87044–658–4 (regular edition)

ISBN 0–87044–663–0 (library edition)

Illustrations Credits

Stephen Frink/WaterHouse (cover, 3 bottom, 100 insets); Florida Division of Historical Resources (1); Bill Curtsinger (2 top, 10 top, 11 top, 13, 18 left, 19 top, 20 bottom, 21 bottom, 35 right bottom, 40 left, 41 bottom); H. Edward Kim (2 center, 36); NATIONAL GEOGRAPHIC Photographer Victor R. Boswell, Jr. (2 bottom, 46–47 bottom, 48, 49 bottom); Pam Smith O'Hara (3 top, 44 top, 45, 68 bottom); National Park Service (3 center); Barbara Gibson (4 top, 10–11 bottom, 12 left, 14–15, 23 top, 28 top, 30–31, 32 bottom, 34 bottom, 41 top, 42–43, 50–51 bottom, 52–53, 60–61 flag, 62–63, 80–81, 94–95, 96–97, 102–104); NATIONAL GEOGRAPHIC Photographer Emory K. Kristof (4 bottom, 68 left); Al Giddings/Ocean Images, Inc. (5, 99).

Michael A. Hampshire (6–7); Louis S. Glanzman (8–9, 24–25, 38–39, 46–47 top, 56–57, 66–67, 71 top, 74–75, 78–79, 86–87, 88–89); Donald Frey (12 right, 16 top right, 17 center); William M. Murray (16 bottom left); Jacques Ostier/Louvre, Paris (16 bottom right); National Archaeological Museum of Athens (16–17); Susan Womer Katzev (17 top); Soprintendenza Archeologica di Siracusa (17 bottom); J. Robert Teringo, N.G.S. Staff (18–19, 20–21 top).

A. Solazzi (22 top, 23 bottom); Soprintendenza Archeologica del Lazio (22 bottom); © University Museum of National Antiquities, Oslo, Norway (26 top left); N.G.S. Photographer Victor R. Boswell, Jr., and Milton A. Ford/Bayeux Tapestry, with special authorization of the Ville de Bayeux (26–27 bottom); reproduced by courtesy of the Trustees of the British Museum (27 top, 71 bottom); The Viking Ship Museum, Roskilde, Denmark (28 bottom left); Ted Spiegel (28 bottom right, 29); Jonathan Blair (32 top left, 33); Robin Piercy/Institute of Nautical Archaeology (32 top right, 34–35); Tomoyuki Narashima (37, 54–55, 92–93).

AB Nordbok, Gothenburg, Sweden, from Conrad Gesner's *Historiae Animalium Liber IV*, Zurich, 1558 (40 bottom); Richard Schlecht (44 bottom, 91 right); The Mary Rose Trust (49 top); Jacob Forsell (50 top, 51 top, 54 top); Vasa Museum, Stockholm: Winfield Parks (51 bottom); Max Lewold and W. E. Roscher (53 left inset); Ira Block (53 right inset, 54 bottom); D. D. Denton/Institute of Nautical Archaeology (58 top); Luis Marden, N.G.S. Staff (58 bottom); Doug Kesling/Institute of Nautical Archaeology (59); Institute of Nautical Archaeology (59 inset).

Mary Evans Picture Library (60); Rare Books & Manuscripts Division/The New York Public Library/Astor, Lenox & Tilden Foundations (61 left); NATIONAL GEOGRAPHIC Photographer Bruce Dale (61 top right); Driscoll Piracy Collection/Wichita Public Library (61 bottom); woodcut by Oronce Fine (1515), from Geo. Peurbach's *Theoricarum Novarum Textus*, printed with permission of The Holland Press Limited, London, from Rodney W. Shirley's *The Mapping of the World: Early Printed World Maps 1472–1700* (64 top); courtesy of the Texas Antiquities Committee and the Corpus Christi Museum (64 right, 65 bottom, 67 top, 67 bottom right).

The University of Texas Institute of Texan Cultures (64 center, 65 center, 67 bottom left); courtesy of the Texas Antiquities Committee (65 top); © David Doubilet (68 top, 69 right, 100–101); David L. Arnold, N.G.S. Staff (69 left); Peter Marsden (70 left); Ben & Lynn Cropp (70 bottom); courtesy of the Freer Gallery of Art, Smithsonian Institution, Washington, D. C./Accession Nos. 65.22 and 65.23 (72–73); NATIONAL GEOGRAPHIC Photographer Bates Littlehales (72 bottom left, 77 top, 79 bottom); photo: Stenuit–Grasp (72 bottom right); Richard Keen (73 bottom).

Marie–Louise Coulson (76 top, 76–77, 78 bottom); courtesy of the William L. Clements Library, University of Michigan (76 bottom); N.G.S. photograph by Emory K. Kristof/© Hamilton–Scourge Foundation (82 top, 83); Canada Centre Inland Waters, Environment Canada, Burlington, Ontario, (82 bottom); John Fulton (84 top left and bottom); U. S. Fish & Wildlife Service (84 top right); David C. Peterson (85, 87); © 1986 Woods Hole Oceanographic Institution (90, 91 left, 95 top).

Ulster Folk and Transport Museum Photographic Archive—Harland & Wolff Collection (92 top); The Cork Examiner Publications Ltd. (92 bottom left); Illustrated London News Picture Library (92 bottom right); photography Ron & Valerie Taylor (98 top); John Hamilton (98 bottom); Marvin J. Fryer (mechanicals).

PAGES 102–103 AND ABOVE: Shown to scale, 16 featured ships pass in review.

COVER: A diver explores a fishing vessel that sank in 1981 in waters of the British Virgin Islands.

HIDDEN TREASURES
OF THE SEA

PUBLISHED BY
THE NATIONAL GEOGRAPHIC SOCIETY
WASHINGTON, D. C.

Gilbert M. Grosvenor, *President and Chairman of the Board*
Melvin M. Payne, Thomas W. McKnew, *Chairmen Emeritus*
Owen R. Anderson, *Executive Vice President*
Robert L. Breeden, *Senior Vice President, Publications and Educational Media*

PRODUCED BY THE SPECIAL PUBLICATIONS AND SCHOOL SERVICES DIVISION
Donald J. Crump, *Director*
Philip B. Silcott, *Associate Director*
Bonnie S. Lawrence, *Assistant Director*

BOOKS FOR WORLD EXPLORERS
Pat Robbins, *Editor*
Ralph Gray, *Editor Emeritus*
Ursula Perrin Vosseler, *Art Director*
Margaret McKelway, *Associate Editor*

STAFF FOR *HIDDEN TREASURES OF THE SEA*
Ross Bankson, *Managing Editor*
Marianne R. Koszorus, *Art Director*
Jennifer A. Kirkpatrick, *Senior Researcher and Project Editor*
Alison Wilbur Eskildsen, *Picture Editor*
Jacqueline N. Thompson, *Researcher*
K. M. Kostyal (Introduction, chapter 6), Ann Di Fiore (chapters 1, 2), Sharon L. Barry (chapters 3, 4, 5), *Writers*
Mary Elizabeth House, *Research Assistant and Editorial Assistant*
Sandra F. Lotterman, *Editorial Assistant*
Janet A. Dustin, Jennie H. Proctor, Karen L. O'Brien, *Illustrations Assistants*
Joseph F. Ochlak, *Map Editor*
Katherine D. Jones, *Picture-Editing Intern;* Kathryn N. Adams, *Editorial Intern;* Regina Rodrigues, Heidi Salter, *Design Interns*

ENGRAVING, PRINTING, AND PRODUCT MANUFACTURE
George V. White, *Manager;* Vincent P. Ryan, *Assistant Manager;* David V. Showers, *Production Manager;* Lewis R. Bassford, *Production Project Manager;* Timothy H. Ewing, *Senior Production Assistant;* Carol R. Curtis, *Senior Production Staff Assistant*

STAFF ASSISTANTS: Betsy Ellison, Kaylene F. Kahler, Eliza C. Morton, Nancy J. White

INDEX: James B. Enzinna

Composition for HIDDEN TREASURES OF THE SEA by the Typographic section of National Geographic Production Services, Pre-Press Division and York Graphic Services, York, Pa. Printed and bound by Holladay–Tyler Printing Corp., Rockville, Md. Film preparation by Catharine Cooke Studio, Inc., New York, N.Y. Color separations by Lincoln Graphics, Inc., Cherry Hill, N.J. Cover printed by Federated Lithographers–Printers, Inc., Providence, R.I. Teacher's Guide printed by McCollum Press, Inc., Rockville, Md.